W9-CBF-158
YOU JUST SAVED
5% AT
Bookstore
NEW
PRICE
USED
PRICE
M C B

The Houghton Mifflin Series in Statistics
under the Editorship of Herman Chernoff

LEO BREIMAN
Probability and Stochastic Processes: With a View Toward Applications
Statistics: With a View Toward Applications

PAUL G. HOEL, SIDNEY C. PORT, AND CHARLES J. STONE
Introduction to Probability Theory
Introduction to Statistical Theory
Introduction to Stochastic Processes

PAUL F. LAZARSFELD AND NEIL W. HENRY
Latent Structure Analysis

GOTTFRIED E. NOETHER
Introduction to Statistics—A Fresh Approach

Y. S. CHOW, HERBERT ROBBINS, AND DAVID SIEGMUND
Great Expectations: The Theory of Optimal Stopping

I. RICHARD SAVAGE
Statistics: Uncertainty and Behavior

Introduction to Probability Theory

Paul G. Hoel
Sidney C. Port
Charles J. Stone
University of California, Los Angeles

Tang Ka Keung Fall 83

HOUGHTON MIFFLIN COMPANY BOSTON
Atlanta Dallas Geneva, Illinois Hopewell, New Jersey
Palo Alto London

PRINTED IN THE U.S.A.

LIBRARY OF CONGRESS CATALOG CARD NUMBER: 74-136173

ISBN: 0-395-04636-X

General Preface

This three-volume series grew out of a three-quarter course in probability, statistics, and stochastic processes taught for a number of years at UCLA. We felt a need for a series of books that would treat these subjects in a way that is well coordinated, but which would also give adequate emphasis to each subject as being interesting and useful on its own merits.

The first volume, *Introduction to Probability Theory*, presents the fundamental ideas of probability theory and also prepares the student both for courses in statistics and for further study in probability theory, including stochastic processes.

The second volume, *Introduction to Statistical Theory*, develops the basic theory of mathematical statistics in a systematic, unified manner. Together, the first two volumes contain the material that is often covered in a two-semester course in mathematical statistics.

The third volume, *Introduction to Stochastic Processes*, treats Markov chains, Poisson processes, birth and death processes, Gaussian processes, Brownian motion, and processes defined in terms of Brownian motion by means of elementary stochastic differential equations.

Preface

This volume is intended to serve as a text for a one-quarter or one-semester course in probability theory at the junior-senior level. The material has been designed to give the reader adequate preparation for either a course in statistics or further study in probability theory and stochastic processes. The prerequisite for this volume is a course in elementary calculus that includes multiple integration.

We have endeavored to present only the more important concepts of probability theory. We have attempted to explain these concepts and indicate their usefulness through discussion, examples, and exercises. Sufficient detail has been included in the examples so that the student can be expected to read these on his own, thereby leaving the instructor more time to cover the essential ideas and work a number of exercises in class.

At the conclusion of each chapter there are a large number of exercises, arranged according to the order in which the relevant material was introduced in the text. Some of these exercises are of a routine nature, while others develop ideas introduced in the text a little further or in a slightly different direction. The more difficult problems are supplied with hints. Answers, when not indicated in the problems themselves, are given at the end of the book.

Although most of the subject matter in this volume is essential for further study in probability and statistics, some optional material has been included to provide for greater flexibility. These optional sections are indicated by an asterisk. The material in Section 6.2.2 is needed only for Section 6.6; neither section is required for this volume, but both are needed in *Introduction to Statistical Theory*. The material of Section 6.7 is used only in proving Theorem 1 of Chapter 9 in this volume and Theorem 1 of Chapter 5 in *Introduction to Statistical Theory*. The contents of Chapters 8 and 9 are optional; Chapter 9 does not depend on Chapter 8.

We wish to thank our several colleagues who read over the original manuscript and made suggestions that resulted in significant improvements. We also would like to thank Neil Weiss and Luis Gorostiza for obtaining answers to all the exercises and Mrs. Ruth Goldstein for her excellent job of typing.

Table of Contents

1 Probability Spaces

Probability theory is the branch of mathematics that is concerned with random (or chance) phenomena. It has attracted people to its study both because of its intrinsic interest and its successful applications to many areas within the physical, biological, and social sciences, in engineering, and in the business world.

Many phenomena have the property that their repeated observation under a specified set of conditions invariably leads to the same outcome. For example, if a ball initially at rest is dropped from a height of d feet through an evacuated cylinder, it will invariably fall to the ground in $t = \sqrt{2d/g}$ seconds, where $g = 32 \text{ ft/sec}^2$ is the constant acceleration due to gravity. There are other phenomena whose repeated observation under a specified set of conditions does not always lead to the same outcome. A familiar example of this type is the tossing of a coin. If a coin is tossed 1000 times the occurrences of heads and tails alternate in a seemingly erratic and unpredictable manner. It is such phenomena that we think of as being random and which are the object of our investigation.

At first glance it might seem impossible to make any worthwhile statements about such random phenomena, but this is not so. Experience has shown that many nondeterministic phenomena exhibit a *statistical regularity* that makes them subject to study. This may be illustrated by considering coin tossing again. For any given toss of the coin we can make no nontrivial prediction, but observations show that for a large number of tosses the proportion of heads seems to fluctuate around some fixed number p between 0 and 1 (p being very near 1/2 unless the coin is severely unbalanced). It appears as if the proportion of heads in n tosses would converge to p if we let n approach infinity. We think of this limiting proportion p as the "probability" that the coin will land heads up in a single toss.

More generally the statement that a certain experimental outcome has probability p can be interpreted as meaning that if the experiment is repeated a large number of times, that outcome would be observed "about" $100p$ percent of the time. This interpretation of probabilities is called the relative frequency interpretation. It is very natural in many applications of probability theory to real world problems, especially to those involving the physical sciences, but it often seems quite artificial. How, for example, could we give a relative frequency interpretation to the probability that a given newborn baby will live at least 70 years? Various attempts have been made, none of them totally acceptable, to give alternative interpretations to such probability statements.

For the mathematical theory of probability the interpretation of probabilities is irrelevant, just as in geometry the interpretation of points, lines, and planes is irrelevant. We will use the relative frequency interpretation of probabilities only as an intuitive motivation for the definitions and theorems we will be developing throughout the book.

1.1. Examples of random phenomena

In this section we will discuss two simple examples of random phenomena in order to motivate the formal structure of the theory.

Example 1. A box has s balls, labeled $1, 2, \ldots, s$ but otherwise identical. Consider the following experiment. The balls are mixed up well in the box and a person reaches into the box and draws a ball. The number of the ball is noted and the ball is returned to the box. The outcome of the experiment is the number on the ball selected. About this experiment we can make no nontrivial prediction.

Suppose we repeat the above experiment n times. Let $N_n(k)$ denote the number of times the ball labeled k was drawn during these n trials of the experiment. Assume that we had, say, $s = 3$ balls and $n = 20$ trials. The outcomes of these 20 trials could be described by listing the numbers which appeared in the order they were observed. A typical result might be

$$1, 1, 3, 2, 1, 2, 2, 3, 2, 3, 3, 2, 1, 2, 3, 3, 1, 3, 2, 2,$$

in which case

$$N_{20}(1) = 5, \qquad N_{20}(2) = 8, \qquad \text{and} \qquad N_{20}(3) = 7.$$

The relative frequencies (i.e., proportion of times) of the outcomes 1, 2, and 3 are then

$$\frac{N_{20}(1)}{20} = .25, \qquad \frac{N_{20}(2)}{20} = .40, \qquad \text{and} \qquad \frac{N_{20}(3)}{20} = .35.$$

As the number of trials gets large we would expect the relative frequencies $N_n(1)/n, \ldots, N_n(s)/n$ to settle down to some fixed numbers $p_1, p_2, \ldots, p_s$ (which according to our intuition in this case should all be $1/s$).

By the relative frequency interpretation, the number p_i would be called the probability that the ith ball will be drawn when the experiment is performed once ($i = 1, 2, \ldots, s$).

We will now make a mathematical model of the experiment of drawing a ball from the box. To do this, we first take a set Ω having s points that we place into one-to-one correspondence with the possible outcomes of the experiment. In this correspondence exactly one point of Ω will be associated with the outcome that the ball labeled k is selected. Call that point ω_k. To the point ω_k we associate the number $p_k = 1/s$ and call it the probability of ω_k. We observe at once that $0 \le p_k \le 1$ and that $p_1 + \cdots + p_s = 1$.

Suppose now that in addition to being numbered from 1 to s the first r balls are colored red and the remaining $s - r$ are colored black. We perform the experiment as before, but now we are only interested in the color of the ball drawn and not its number. A moment's thought shows that the relative frequency of red balls drawn among n repetitions of the experiment is just the sum of the relative frequencies $N_n(k)/n$ over those values of k that correspond to a red ball. We would expect, and experience shows, that for large n this relative frequency should settle down to some fixed number. Since for large n the relative frequencies $N_n(k)/n$ are expected to be close to $p_k = 1/s$, we would anticipate that the relative frequency of red balls would be close to r/s. Again experience verifies this. According to the relative frequency interpretation, we would then call r/s the probability of obtaining a red ball.

Let us see how we can reflect this fact in our model. Let A be the subset of Ω consisting of those points ω_k such that ball k is red. Then A has exactly r points. We call A an event. More generally, in this situation we will call any subset B of Ω an event. To say the event B occurs means that the outcome of the experiment is represented by some point in B.

Let A and B be two events. Recall that the union of A and B, $A \cup B$, is the set of all points $\omega \in \Omega$ such that $\omega \in A$ or $\omega \in B$. Now the points in Ω are in correspondence with the outcomes of our experiment. The event A occurs if the experiment yields an outcome that is represented by some point in A, and similarly the event B occurs if the outcome of the experiment is represented by some point in B. The set $A \cup B$ then represents the fact that the event A occurs or the event B occurs. Similarly the intersection $A \cap B$ of A and B consists of all points that are in both A and B. Thus if $\omega \in A \cap B$ then $\omega \in A$ and $\omega \in B$ so $A \cap B$ represents the fact that both the events A and B occur. The complement A^c (or A') of A is the set of points in Ω that are not in A. The event A does not occur if the experiment yields an outcome represented by a point in A^c.

Diagrammatically, if A and B are represented by the indicated regions in Figure 1a, then $A \cup B$, $A \cap B$, and A^c are represented by the shaded regions in Figures 1b, 1c, and 1d, respectively.

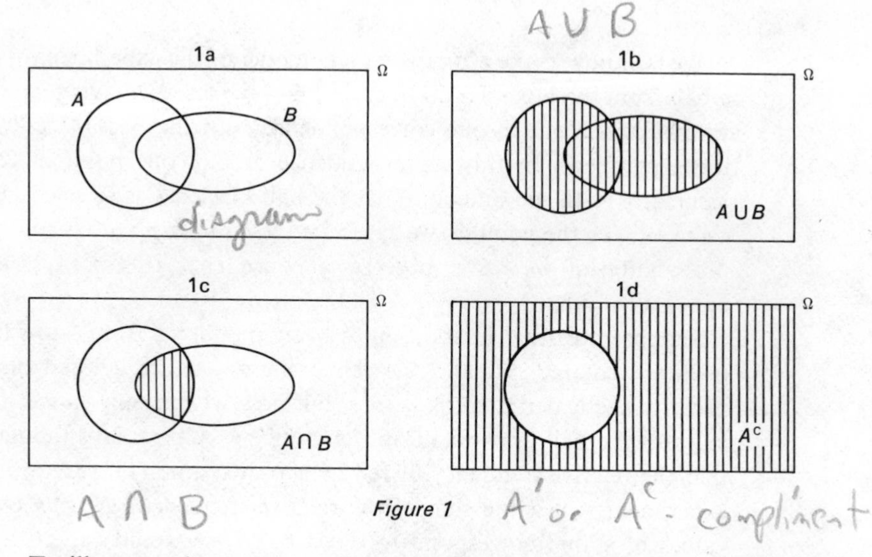

Figure 1

To illustrate these concepts let A be the event "red ball selected" and let B be the event "even-numbered ball selected." Then the union $A \cup B$ is the event that either a red ball or an even-numbered ball was selected. The intersection $A \cap B$ is the event "red even-numbered ball selected." The event A^c occurs if a red ball was not selected.

We now would like to assign probabilities to events. Mathematically, this just means that we associate to each set B a real number. A priori we could do this in an arbitrary way. However, we are restricted if we want these probabilities to reflect the experiment we are trying to model. How should we make this assignment? We have already assigned each point the number s^{-1}. Thus a one-point set $\{\omega\}$ should be assigned the number s^{-1}. Now from our discussion of the relative frequency of the event "drawing a red ball," it seems that we should assign the event A the probability $P(A) = r/s$. More generally, if B is any event we will define $P(B)$ by $P(B) = j/s$ if B has exactly j points. We then observe that

$$P(B) = \sum_{\omega_k \in B} p_k ,$$

where $\sum_{\omega_k \in B} p_k$ means that we sum the numbers p_k over those values of k such that $\omega_k \in B$. From our definition of $P(B)$ it easily follows that the following statements are true. We leave their verification to the reader.

Let $\varnothing$ denote the empty set; then $P(\varnothing) = 0$ and $P(\Omega) = 1$. If A and B are any two disjoint sets, i.e., $A \cap B = \varnothing$, then

$$P(A \cup B) = P(A) + P(B).$$

Example 2. It is known from physical experiments that an isotope of a certain substance is unstable. In the course of time it decays by the emission of neutrons to a stable form. We are interested in the time that it takes an atom of the isotope to decay to its stable form. According to the laws of physics it is impossible to say with certainty when a specified atom of the isotope will decay, but if we observe a large number N of atoms initially, then we can make some accurate predictions about the number of atoms $N(t)$ that have not decayed by time t. In other words we can rather accurately predict the fraction of atoms $N(t)/N$ that have not decayed by time t, but we cannot say which of the atoms will have done so. Since all of the atoms are identical, observing N atoms simultaneously should be equivalent to N repetitions of the same experiment where, in this case, the experiment consists in observing the time that it takes an atom to decay.

Now to a first approximation (which is actually quite accurate) the rate at which the isotope decays at time t is proportional to the number of atoms present at time t, so $N(t)$ is given approximately as the solution of the differential equation

$$\frac{df}{dt} = -\lambda f(t), \qquad f(0) = N,$$

where $\lambda > 0$ is a fixed constant of proportionality. The unique solution of this equation is $f(t) = Ne^{-\lambda t}$, and thus the fraction of atoms that have not decayed by time t is given approximately by $N(t)/N = e^{-\lambda t}$. If $0 \le t_0 \le t_1$, the fraction of atoms that decay in the time interval $[t_0, t_1]$ is $(e^{-\lambda t_0} - e^{-\lambda t_1})$. Consequently, in accordance with the relative frequency interpretation of probability we take $(e^{-\lambda t_0} - e^{-\lambda t_1})$ as the probability that an atom decays between times t_0 and t_1.

To make a mathematical model of this experiment we can try to proceed as in the previous example. First we choose a set Ω that can be put into a one-to-one correspondence with the possible outcomes of the experiment. An outcome in this case is the time that an atom takes to decay. This can be any positive real number, so we take Ω to be the interval $[0, \infty)$ on the real line. From our discussion above it seems reasonable to assign to the interval $[t_0, t_1]$ the probability $(e^{-\lambda t_0} - e^{-\lambda t_1})$. In particular if $t_0 = t_1 = t$ then the interval degenerates to the set $\{t\}$ and the probability assigned to this set is 0.

In our previous example Ω had only finitely many points; however, here Ω has a (noncountable) infinity of points and each point has probability 0. Once again we observe that $P(\Omega) = 1$ and $P(\varnothing) = 0$. Suppose A and B are two disjoint intervals. Then the proportion of atoms that decay in the time interval $A \cup B$ is the sum of the proportion that decay in the time interval A and the proportion that decay in the time interval B. In light of this additivity we demand that in the mathematical model, $A \cup B$ should have probability $P(A) + P(B)$ assigned to it. In other words, in the mathematical model we want

$$P(A \cup B) = P(A) + P(B)$$

whenever A and B are two disjoint intervals.

Preserves linearity of the linear transformation f: = finding the P(A)

1.2. Probability spaces

Our purpose in this section is to develop the formal mathematical structure, called a probability space, that forms the foundation for the mathematical treatment of random phenomena.

Envision some real or imaginary experiment that we are trying to model. The first thing we must do is decide on the possible outcomes of the experiment. It is not too serious if we admit more things into our consideration than can really occur, but we want to make sure that we do not exclude things that might occur. Once we decide on the possible outcomes, we choose a set Ω whose points ω are associated with these outcomes. From the strictly mathematical point of view, however, Ω is just an abstract set of points.

We next take a nonempty collection $\mathscr{A}$ of subsets of Ω that is to represent the collection of "events" to which we wish to assign probabilities. By definition, now, an *event* means a set A in $\mathscr{A}$. The statement *the event A occurs* means that the outcome of our experiment is represented by some point $\omega \in A$. Again, from the strictly mathematical point of view, $\mathscr{A}$ is just a specified collection of subsets of the set Ω. Only sets $A \in \mathscr{A}$, i.e., events, will be assigned probabilities. In our model in Example 1, $\mathscr{A}$ consisted of all subsets of Ω. In the general situation when Ω does not have a finite number of points, as in Example 2, it may not be possible to choose $\mathscr{A}$ in this manner.

The next question is, what should the collection $\mathscr{A}$ be? It is quite reasonable to demand that $\mathscr{A}$ be closed under finite unions and finite intersections of sets in $\mathscr{A}$ as well as under complementation. For example, if A and B are two events, then $A \cup B$ occurs if the outcome of our experiment is either represented by a point in A or a point in B. Clearly, then, if it is going to be meaningful to talk about the probabilities that A and B occur, it should also be meaningful to talk about the probability that either A or B occurs, i.e., that the event $A \cup B$ occurs. Since only sets in $\mathscr{A}$ will be assigned probabilities, we should require that $A \cup B \in \mathscr{A}$ whenever A and B are members of $\mathscr{A}$. Now $A \cap B$ occurs if the outcome of our experiment is represented by some point that is in both A and B. A similar line of reasoning to that used for $A \cup B$ convinces us that we should have $A \cap B \in \mathscr{A}$ whenever $A, B \in \mathscr{A}$. Finally, to say that the event A does not occur is to say that the outcome of our experiment is not represented by a point in A, so that it must be represented by some point in A^c. It would be the height of folly to say that we could talk about the probability of A but not of A^c. Thus we shall demand that whenever A is in $\mathscr{A}$ so is A^c.

We have thus arrived at the conclusion that $\mathscr{A}$ should be a nonempty collection of subsets of Ω having the following properties:

(i) If A is in $\mathscr{A}$ so is A^c.
(ii) If A and B are in $\mathscr{A}$ so are $A \cup B$ and $A \cap B$.

An easy induction argument shows that if $A_1, A_2, \ldots, A_n$ are sets in $\mathscr{A}$ then so are $\bigcup_{i=1}^n A_i$ and $\bigcap_{i=1}^n A_i$. Here we use the shorthand notation

$$\bigcup_{i=1}^{n} A_i = A_1 \cup A_2 \cup \cdots \cup A_n$$

and

$$\bigcap_{i=1}^{n} A_i = A_1 \cap A_2 \cap \cdots \cap A_n.$$

Also, since $A \cap A^c = \emptyset$ and $A \cup A^c = \Omega$, we see that both the empty set $\emptyset$ and the set Ω must be in $\mathscr{A}$.

A nonempty collection of subsets of a given set Ω that is closed under finite set theoretic operations is called a *field of subsets* of Ω. It therefore seems we should demand that $\mathscr{A}$ be a field of subsets. It turns out, however, that for certain mathematical reasons just taking $\mathscr{A}$ to be a field of subsets of Ω is insufficient. What we will actually demand of the collection $\mathscr{A}$ is more stringent. We will demand that $\mathscr{A}$ be closed not only under finite set theoretic operations but under countably infinite set theoretic operations as well. In other words if $\{A_n\}$, $n \geq 1$, is a sequence of sets in $\mathscr{A}$, we will demand that

$$\bigcup_{n=1}^{\infty} A_n \in \mathscr{A} \qquad \text{and} \qquad \bigcap_{n=1}^{\infty} A_n \in \mathscr{A}.$$

Here we are using the shorthand notation

$$\bigcup_{n=1}^{\infty} A_n = A_1 \cup A_2 \cup \cdots$$

to denote the union of all the sets of the sequence, and

$$\bigcap_{n=1}^{\infty} A_n = A_1 \cap A_2 \cap \cdots$$

to denote the intersection of all the sets of the sequence. A collection of subsets of a given set Ω that is closed under countable set theory operations is called a σ-field of subsets of Ω. (The σ is put in to distinguish such a collection from a field of subsets.) More formally we have the following:

Definition 1 *A nonempty collection of subsets $\mathscr{A}$ of a set Ω is called a σ-field of subsets of Ω provided that the following two properties hold:*

(i) *If A is in $\mathscr{A}$, then A^c is also in $\mathscr{A}$.*
(ii) *If A_n is in $\mathscr{A}$, $n = 1, 2, \ldots$, then $\bigcup_{n=1}^{\infty} A_n$ and $\bigcap_{n=1}^{\infty} A_n$ are both in $\mathscr{A}$.*

We now come to the assignment of probabilities to events. As was made clear in the examples of the preceding section, the probability of an event is a nonnegative real number. For an event A, let $P(A)$ denote this number. Then $0 \leq P(A) \leq 1$. The set Ω representing every possible outcome should, of course, be assigned the number 1, so $P(\Omega) = 1$. In our discussion of Example 1 we showed that the probability of events satisfies the property that if A and B are any two disjoint events then $P(A \cup B) = P(A) + P(B)$. Similarly, in Example 2 we showed that if A and B are two disjoint intervals, then we should also require that

$$P(A \cup B) = P(A) + P(B).$$

It now seems reasonable in general to demand that if A and B are disjoint events then $P(A \cup B) = P(A) + P(B)$. By induction, it would then follow that if $A_1, A_2, \ldots, A_n$ are any n mutually disjoint sets (that is, if $A_i \cap A_j = \varnothing$ whenever $i \neq j$), then

$$P\left(\bigcup_{i=1}^{n} A_i\right) = \sum_{i=1}^{n} P(A_i).$$

Actually, again for mathematical reasons, we will in fact demand that this additivity property hold for countable collections of disjoint events.

Definition 2 *A probability measure P on a σ-field of subsets $\mathscr{A}$ of a set Ω is a real-valued function having domain $\mathscr{A}$ satisfying the following properties:*

(i) $P(\Omega) = 1$.

(ii) $P(A) \geq 0$ *for all* $A \in \mathscr{A}$.

(iii) *If* A_n, $n = 1, 2, 3, \ldots$, *are mutually disjoint sets in* $\mathscr{A}$, *then*

$$P\left(\bigcup_{n=1}^{\infty} A_n\right) = \sum_{n=1}^{\infty} P(A_n).$$

A probability space, denoted by $(\Omega, \mathscr{A}, P)$, is a set Ω, a σ-field of subsets $\mathscr{A}$, and a probability measure P defined on $\mathscr{A}$.

It is quite easy to find a probability space that corresponds to the experiment of drawing a ball from a box. In essence it was already given in our discussion of that experiment. We simply take Ω to be a finite set having s points, $\mathscr{A}$ to be the collection of all subsets of Ω, and P to be the probability measure that assigns to A the probability $P(A) = j/s$ if A has exactly j points.

Let us now consider the probability space associated with the isotope disintegration experiment (Example 2). Here it is certainly clear that $\Omega = [0, \infty)$, but it is not obvious what $\mathscr{A}$ and P should be. Indeed, as we will indicate below, this is by no means a trivial problem, and one that in all its ramifications depends on some deep properties of set theory that are beyond the scope of this book.

One thing however is clear; whatever $\mathscr{A}$ and P are chosen to be, $\mathscr{A}$ must contain all intervals, and P must assign probability $(e^{-\lambda t_0} - e^{-\lambda t_1})$ to the interval $[t_0, t_1]$ if we want the probability space we are constructing to reflect the physical situation. The problem of constructing the space now becomes the following purely mathematical one. Is there a σ-field $\mathscr{A}$ that contains all intervals as members and a probability measure P defined on $\mathscr{A}$ that assigns the desired probability $P(A)$ to the interval A? Questions of this type are in the province of a branch of advanced mathematics called measure theory and cannot be dealt with at the level of this book. Results from measure theory show that the answer to this particular question and others of a similar nature is yes, so that such constructions are always possible.

We will not dwell on the construction of probability spaces in general. The mathematical theory of probability begins with an abstract probability space and develops the theory using the probability space as a basis of operation. Aside from forming a foundation for precisely defining other concepts in the theory, the probability space itself plays very little role in the further development of the subject. Auxiliary quantities (especially random variables, a concept taken up in Chapter 3) quickly become the dominant theme of the theory and the probability space itself fades into the background.

We will conclude our discussion of probability spaces by constructing an important class of probability spaces, called *uniform probability spaces*.

Some of the oldest problems in probability involve the idea of picking a point "at random" from a set S. Our intuitive ideas on this notion show us that if A and B are two subsets having the same "size" then the chance of picking a point from A should be the same as from B. If S has only finitely many points we can measure the "size" of a set by its cardinality. Two sets are then of the same "size" if they have the same number of points. It is quite easy to make a probability space corresponding to the experiment of picking a point at random from a set S having a finite number s of points. We take $\Omega = S$ and $\mathscr{A}$ to be all subsets of S, and assign to the set A the probability $P(A) = j/s$ if A is a set having exactly j points. Such a probability space is called a *symmetric probability space* because each one-point set carries the same probability s^{-1}. We shall return to the study of such spaces in Chapter 2.

Suppose now that S is the interval $[a, b]$ on the real line where $-\infty < a < b < +\infty$. It seems reasonable in this case to measure the "size" of a subset A of $[a, b]$ by its length. Two sets are then of the same size if they have the same length. We will denote the length of a set A by $|A|$.

To construct a probability space for the experiment of "choosing a point at random from S," we proceed in a manner similar to that used for the isotope experiment. We take $\Omega = S$, and appeal to the results of measure theory that show that there is a σ-field $\mathscr{A}$ of subsets of S, and a probability measure P defined on $\mathscr{A}$ such that $P(A) = |A|/|S|$ whenever A is an interval.

More generally, let S be any subset of r-dimensional Euclidean space having finite, nonzero r-dimensional volume. For a subset A of S denote the volume of A by $|A|$. There is then a σ-field $\mathscr{A}$ of subsets of S that contains all the subsets of S that have volume assigned to them as in calculus, and a probability measure P defined on $\mathscr{A}$ such that $P(A) = |A|/|S|$ for any such set A. We will call any such probability space, denoted by $(S, \mathscr{A}, P)$, a *uniform probability space.*

1.3. Properties of probabilities

In this section we will derive some additional properties of a probability measure P that follow from the definition of a probability measure. These properties will be used constantly throughout the remainder of the book. We assume that we are given some probability space $(\Omega, \mathscr{A}, P)$ and that all sets under discussion are events, i.e., members of $\mathscr{A}$.

For any set A, $A \cup A^c = \Omega$ and thus for any two sets A and B we have the decomposition of B:

$$B = \Omega \cap B = (A \cup A^c) \cap B = (A \cap B) \cup (A^c \cap B). \tag{1}$$

Since $A \cap B$ and $A^c \cap B$ are disjoint, we see from (iii) of Definition 2 that

$$P(B) = P(A \cap B) + P(A^c \cap B). \tag{2}$$

By setting $B = \Omega$ and recalling that $P(\Omega) = 1$, we conclude from (2) that

$$P(A^c) = 1 - P(A). \tag{3}$$

In particular $P(\varnothing) = 1 - P(\Omega)$, so that

$$P(\varnothing) = 0. \tag{4}$$

As a second application of (2) suppose that $A \subset B$. Then $A \cap B = A$ and hence

$$P(B) = P(A) + P(A^c \cap B) \qquad \text{if} \quad A \subset B. \tag{5}$$

Since $P(A^c \cap B) \geq 0$ by (ii), we see from (5) that

$$P(B) \geq P(A) \qquad \text{if} \quad A \subset B. \tag{6}$$

De Morgan's laws state that if $\{A_n\}$, $n \geq 1$, is any sequence of sets, then

$$\left(\bigcup_n A_n\right)^c = \left(\bigcap_n A_n^c\right) \tag{7}$$

and

$$(8) \qquad \left(\bigcap_n A_n\right)^c = \left(\bigcup_n A_n^c\right).$$

To see that (7) holds, observe that $\omega \in (\bigcup_{n \geq 1} A_n)^c$ if and only if $\omega \notin A_n$ for any n; that is, $\omega \in A_n^c$ for all $n \geq 1$, or equivalently, $\omega \in \bigcap_n A_n^c$. To establish (8) we apply (7) to $\{A_n^c\}$, obtaining

$$\left(\bigcup_n A_n^c\right)^c = \bigcap_n A_n,$$

and by taking complements, we see that

$$\bigcup_n A_n^c = \left(\bigcap_n A_n\right)^c.$$

A useful relation that follows from (7) and (3) is

$$(9) \qquad P\left(\bigcup_n A_n\right) = 1 - P\left(\bigcap_n A_n^c\right).$$

Now $\bigcup_n A_n$ is the event that at least one of the events A_n occurs, while $\bigcap_n A_n^c$ is the event that none of these events occur. In words, (9) asserts that the probability that at least one of the events A_n will occur is 1 minus the probability that none of the events A_n will occur. The advantage of (9) is that in some instances it is easier to compute $P(\bigcap_n A_n^c)$ than to compute $P(\bigcup_n A_n)$. [Note that since the events A_n are not necessarily disjoint it is not true that $P(\bigcup_n A_n) = \sum_n P(A_n)$.] The use of (9) is nicely illustrated by means of the following.

Example 3. Suppose three perfectly balanced and identical coins are tossed. Find the probability that at least one of them lands heads.

There are eight possible outcomes of this experiment as follows:

Coin 1	H	H	H	H	T	T	T	T
Coin 2	H	H	T	T	H	H	T	T
Coin 3	H	T	H	T	H	T	H	T

Our intuitive notions suggest that each of these eight outcomes should have the probability 1/8. Let A_1 be the event that the first coin lands heads, A_2 the event that the second coin lands heads, and A_3 the event that the third coin lands heads. The problem asks us to compute $P(A_1 \cup A_2 \cup A_3)$. Now $A_1^c \cap A_2^c \cap A_3^c = \{\mathrm{T, T, T}\}$ and thus

$$P(A_1^c \cap A_2^c \cap A_3^c) = 1/8;$$

hence (9) implies that

$$P(A_1 \cup A_2 \cup A_3) = 1 - P(A_1^c \cap A_2^c \cap A_3^c) = 7/8.$$

Our basic postulate (iii) on probability measures tells us that for disjoint sets A and B, $P(A \cup B) = P(A) + P(B)$. If A and B are not necessarily disjoint, then

$$P(A \cup B) = P(A) + P(B) - P(A \cap B) \tag{10}$$

and consequently

$$P(A \cup B) \leq P(A) + P(B). \tag{11}$$

To see that (10) is true observe that the sets $A \cap B^c$, $A \cap B$, and $A^c \cap B$ are mutually disjoint and their union is just $A \cup B$ (see Figure 2). Thus

$$P(A \cup B) = P(A \cap B^c) + P(A^c \cap B) + P(A \cap B). \tag{12}$$

By (2), however,

$$P(A \cap B^c) = P(A) - P(A \cap B)$$

and

$$P(A^c \cap B) = P(B) - P(A \cap B).$$

By substituting these expressions into (12), we obtain (10).

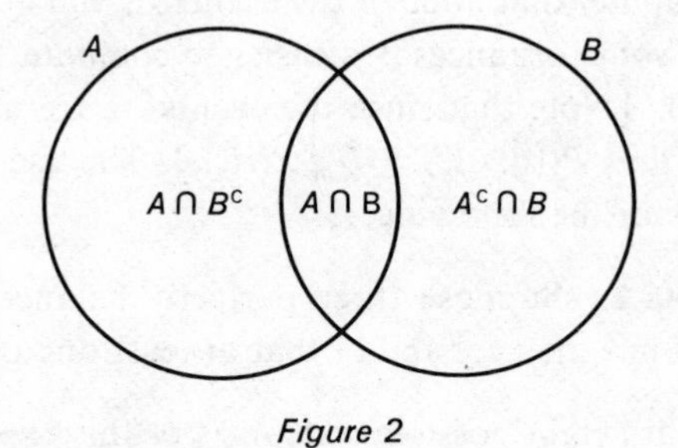

Figure 2

Equations (10) and (11) extend to any finite number of sets. The analogue of the exact formula (10) is a bit complicated and will be discussed in Chapter 2. Inequality (11), however, can easily be extended by induction to yield

$$P(A_1 \cup A_2 \cup \cdots \cup A_n) \leq \sum_{i=1}^{n} P(A_i). \tag{13}$$

To prove this, observe that if $n \geq 2$, then by (11)

$$\begin{aligned} P(A_1 \cup \cdots \cup A_n) &= P((A_1 \cup \cdots \cup A_{n-1}) \cup A_n) \\ &\leq P(A_1 \cup \cdots \cup A_{n-1}) + P(A_n). \end{aligned}$$

Hence if (13) holds for $n - 1$ sets, it holds for n sets. Since (13) clearly holds for $n = 1$, the result is proved by induction.

So far we have used only the fact that a probability measure is finitely additive. Our next result will use the countable additivity.

Theorem 1 *Let A_n, $n \geq 1$, be events.*

(i) *If $A_1 \subset A_2 \subset \cdots$ and $A = \bigcup_{n=1}^{\infty} A_n$, then*

$$\lim_{n \to \infty} P(A_n) = P(A). \tag{14}$$

(ii) *If $A_1 \supset A_2 \supset \cdots$ and $A = \bigcap_{n=1}^{\infty} A_n$, then (14) again holds.*

Proof of (i). Suppose $A_1 \subset A_2 \subset \cdots$ and $A = \bigcup_{n=1}^{\infty} A_n$. Set $B_1 = A_1$, and for each $n \geq 2$, let B_n denote those points which are in A_n but not in A_{n-1}, i.e., $B_n = A_n \cap A_{n-1}^c$. A point ω is in B_n if and only if ω is in A and A_n is the first set in the sequence $A_1, A_2, \ldots$ containing ω. By definition, the sets B_n are disjoint,

$$A_n = \bigcup_{i=1}^{n} B_i,$$

and

$$A = \bigcup_{i=1}^{\infty} B_i.$$

Consequently,

$$P(A_n) = \sum_{i=1}^{n} P(B_i)$$

and

$$P(A) = \sum_{i=1}^{\infty} P(B_i).$$

Now

$$\lim_{n \to \infty} \sum_{i=1}^{n} P(B_i) = \sum_{i=1}^{\infty} P(B_i) \tag{15}$$

by the definition of the sum of an infinite series. It follows from (15) that

$$\lim_{n \to \infty} P(A_n) = \lim_{n \to \infty} \sum_{i=1}^{n} P(B_i)$$

$$= \sum_{i=1}^{\infty} P(B_i) = P(A),$$

so that (14) holds.

Proof of (ii). Suppose $A_1 \supset A_2 \supset \cdots$ and $A = \bigcap_{n=1}^{\infty} A_n$. Then $A_1^c \subset A_2^c \subset \cdots$ and by (8)

$$A^c = \bigcup_{n=1}^{\infty} A_n^c.$$

Thus by (i) of the theorem

$$\lim_{n \to \infty} P(A_n^c) = P(A^c). \tag{16}$$

Since $P(A_n^c) = 1 - P(A_n)$ and $P(A^c) = 1 - P(A)$, it follows from (16) that

$$\begin{aligned}\lim_{n\to\infty} P(A_n) &= \lim_{n\to\infty} (1 - P(A_n^c))\\ &= 1 - \lim_{n\to\infty} P(A_n^c)\\ &= 1 - P(A^c) = P(A),\end{aligned}$$

and again (14) holds. ∎

1.4. Conditional probability

Suppose a box contains r red balls labeled $1, 2, \ldots, r$ and b black balls labeled $1, 2, \ldots, b$. Assume that the probability of drawing any particular ball is $(b + r)^{-1}$. If the ball drawn from the box is known to be red, what is the probability that it was the red ball labeled 1? Another way of stating this problem is as follows. Let A be the event that the selected ball was red, and let B be the event that the selected ball was labeled 1. The problem is then to determine the "conditional" probability that the event B occurred, given that the event A occurred. This problem cannot be solved until a precise definition of the conditional probability of one event given another is available. This definition is as follows:

Definition 3 *Let A and B be two events such that $P(A) > 0$. Then the conditional probability of B given A, written $P(B \mid A)$, is defined to be*

$$P(B \mid A) = \frac{P(B \cap A)}{P(A)}. \tag{17}$$

If $P(A) = 0$ the conditional probability of B given A is undefined.

The above definition is quite easy to motivate by the relative frequency interpretation of probabilities. Consider an experiment that is repeated a large number of times. Let the number of times the events A, B, and $A \cap B$ occur in n trials of the experiment be denoted by $N_n(A)$, $N_n(B)$, and $N_n(A \cap B)$, respectively. For n large we expect that $N_n(A)/n$, $N_n(B)/n$, and $N_n(A \cap B)/n$ should be close to $P(A)$, $P(B)$, and $P(A \cap B)$, respectively. If now we just record those experiments in which A occurs then we have $N_n(A)$ trials in which the event B occurs $N_n(A \cap B)$ times. Thus the proportion of times that B occurs among these $N_n(A)$ experiments is $N_n(A \cap B)/N_n(A)$. But

$$\frac{N_n(A \cap B)}{N_n(A)} = \frac{N_n(A \cap B)/n}{N_n(A)/n}$$

and thus for large values of n this fraction should be close to

$$P(A \cap B)/P(A).$$

As a first example of the use of (17) we will solve the problem posed at the start of this section. Since the set Ω has $b + r$ points each of which carries the probability $(b + r)^{-1}$, we see that $P(A) = r(b + r)^{-1}$ and $P(A \cap B) = (b + r)^{-1}$. Thus

$$P(B \mid A) = \frac{1}{r}.$$

This should be compared with the "unconditional" probability of B, namely $P(B) = 2(b + r)^{-1}$.

Example 4. Suppose two identical and perfectly balanced coins are tossed once.

(a) Find the conditional probability that both coins show a head given that the first shows a head.

(b) Find the conditional probability that both are heads given that at least one of them is a head.

To solve these problems we let the probability space Ω consist of the four points HH, HT, TH, TT, each carrying probability 1/4. Let A be the event that the first coin results in heads and let B be the event that the second coin results in heads. To solve (a) we compute

$$P(A \cap B \mid A) = P(A \cap B)/P(A) = (1/4)/(1/2) = 1/2.$$

To answer (b) we compute

$$P(A \cap B \mid A \cup B) = P(A \cap B)/P(A \cup B) = (1/4)/(3/4) = 1/3.$$

In the above two examples the probability space was specified, and we used (17) to compute various conditional probabilities. In many problems however, we actually proceed in the opposite direction. We are given in advance what we want some conditional probabilities to be, and we use this information to compute the probability measure on Ω. A typical example of this situation is the following.

Example 5. Suppose that the population of a certain city is 40% male and 60% female. Suppose also that 50% of the males and 30% of the females smoke. Find the probability that a smoker is male.

Let M denote the event that a person selected is a male and let F denote the event that the person selected is a female. Also let S denote the event that the person selected smokes and let N denote the event that he does not smoke. The given information can be expressed in the form $P(S \mid M) = .5$,

$P(S \mid F) = .3$, $P(M) = .4$, and $P(F) = .6$. The problem is to compute $P(M \mid S)$. By (17),

$$P(M \mid S) = \frac{P(M \cap S)}{P(S)}.$$

Now $P(M \cap S) = P(M)P(S \mid M) = (.4)(.5) = .20$, so the numerator can be computed in terms of the given probabilities. Since S is the union of the two disjoint sets $S \cap M$ and $S \cap F$, it follows that

$$P(S) = P(S \cap M) + P(S \cap F).$$

Since

$$P(S \cap F) = P(F)P(S \mid F) = (.6)(.3) = .18,$$

we see that

$$P(S) = .20 + .18 = .38.$$

Thus

$$P(M \mid S) = \frac{.20}{.38} \approx .53.$$

The reader will notice that the probability space, as such, was never explicitly mentioned. This problem and others of a similar type are solved simply by using the given data and the rules of computing probabilities given in Section 3 to compute the requested probabilities.

It is quite easy to construct a probability space for the above example. Take the set Ω to consist of the four points SM, SF, NM, and NF that are, respectively, the unique points in the sets $S \cap M$, $S \cap F$, $N \cap M$, and $N \cap F$. The probabilities attached to these four points are not directly specified, but are to be computed so that the events $P(S \mid M)$, $P(S \mid F)$, $P(M)$, and $P(F)$ have the prescribed probabilities. We have already found that $P(S \cap M) = .20$ and $P(S \cap F) = .18$. We leave it as an exercise to compute the probabilities attached to the other two points.

The problem discussed in this example is a special case of the following general situation. Suppose $A_1, A_2, \ldots, A_n$ are n mutually disjoint events with union Ω. Let B be an event such that $P(B) > 0$ and suppose $P(B \mid A_k)$ and $P(A_k)$ are specified for $1 \le k \le n$. What is $P(A_i \mid B)$? To solve this problem note that the A_k are disjoint sets with union Ω and consequently

$$B = B \cap \left(\bigcup_{k=1}^{n} A_k \right) = \bigcup_{k=1}^{n} (B \cap A_k).$$

Thus

$$P(B) = \sum_{k=1}^{n} P(B \cap A_k).$$

But

$$P(B \cap A_k) = P(A_k)P(B \mid A_k),$$

so we can write

$$(18) \qquad P(A_i \mid B) = \frac{P(A_i \cap B)}{P(B)} = \frac{P(A_i)P(B \mid A_i)}{\sum_{k=1}^{n} P(A_k)P(B \mid A_k)}.$$

This formula, called *Bayes' rule*, finds frequent application. One way of looking at the result in (18) is as follows. Suppose we think of the events A_k as being the possible "causes" of the observable event B. Then $P(A_i \mid B)$ is the probability that the event A_i was the "cause" of B given that B occurs. Bayes' rule also forms the basis of a statistical method called Bayesian procedures that will be discussed in Volume II, *Introduction to Statistical Theory*.

As an illustration of the use of Bayes' rule we consider the following (somewhat classical) problem.

Example 6. Suppose there are three chests each having two drawers. The first chest has a gold coin in each drawer, the second chest has a gold coin in one drawer and a silver coin in the other drawer, and the third chest has a silver coin in each drawer. A chest is chosen at random and a drawer opened. If the drawer contains a gold coin, what is the probability that the other drawer also contains a gold coin? We ask the reader to pause and guess what the answer is before reading the solution. Often in this problem the erroneous answer of 1/2 is given.

This problem is easily and correctly solved using Bayes' rule once the description is deciphered. We can think of a probability space being constructed in which the events A_1, A_2, and A_3 correspond, respectively, to the first, second, and third chest being selected. These events are disjoint and their union is the whole space Ω since exactly one chest is selected. Moreover, it is presumably intended that the three chests are equally likely of being chosen so that $P(A_i) = 1/3$, $i = 1, 2, 3$. Let B be the event that the coin observed was gold. Then, from the composition of the chests it is clear that

$$P(B \mid A_1) = 1, \qquad P(B \mid A_2) = 1/2, \qquad \text{and} \qquad P(B \mid A_3) = 0.$$

The problem asks for the probability that the second drawer has a gold coin given that there was a gold coin in the first. This can only happen if the chest selected was the first, so the problem is equivalent to finding $P(A_1 \mid B)$. We now can apply Bayes' rule (18) to compute the answer, which is 2/3. We leave it to the reader as an exercise to compute the probability that the second drawer has a silver coin given that the first drawer had a gold coin.

For our next example we consider a simple probability scheme due to Polya.

Example 7. Polya's urn scheme. Suppose an urn has r red balls and b black balls. A ball is drawn and its color noted. Then it together with $c > 0$ balls of the same color as the drawn ball are added to the urn. The procedure is repeated $n - 1$ additional times so that the total number of drawings made from the urn is n.

Let R_j, $1 \leq j \leq n$, denote the event that the jth ball drawn is red and let B_j, $1 \leq j \leq n$, denote the event that the jth ball drawn is black. Of course, for each j, R_j and B_j are disjoint. At the kth draw there are $b + r + (k - 1)c$ balls in the urn and we assume that the probability of drawing any particular ball is $(b + r + (k - 1)c)^{-1}$. To compute $P(R_1 \cap R_2)$ we write

$$P(R_1 \cap R_2) = P(R_1)P(R_2 \mid R_1).$$

Now

$$P(R_1) = \frac{r}{b + r}, \qquad P(R_2 \mid R_1) = \frac{r + c}{b + r + c},$$

and thus

$$P(R_1 \cap R_2) = \left(\frac{r}{b + r}\right)\left(\frac{r + c}{b + r + c}\right).$$

Similarly

$$P(B_1 \cap R_2) = \left(\frac{b}{b + r}\right)\left(\frac{r}{b + r + c}\right)$$

and thus

$$\begin{aligned} P(R_2) &= P(R_1 \cap R_2) + P(B_1 \cap R_2) \\ &= \left(\frac{r}{b + r}\right)\left(\frac{r + c}{b + r + c}\right) + \left(\frac{b}{b + r}\right)\left(\frac{r}{b + r + c}\right) \\ &= \frac{r}{b + r}. \end{aligned}$$

Consequently, $P(R_2) = P(R_1)$. Since

$$P(B_2) = 1 - P(R_2) = \frac{b}{b + r},$$

$P(B_2) = P(B_1)$. Further properties of the Polya scheme will be developed in the exercises.

1.5. Independence

Consider a box having four distinct balls and an experiment consisting of selecting a ball from the box. We assume that the balls are equally likely to be drawn. Let $\Omega = \{1, 2, 3, 4\}$ and assign probability 1/4 to each point.

Let A and B be two events. For some choices of A and B, knowledge that A occurs increases the odds that B occurs. For example, if $A = \{1, 2\}$ and $B = \{1\}$, then $P(A) = 1/2$, $P(B) = 1/4$, and $P(A \cap B) = 1/4$. Consequently, $P(B \mid A) = 1/2$, which is greater than $P(B)$. On the other hand, for other choices of A and B, knowledge that A occurs decreases the odds that B will occur. For example, if $A = \{1, 2, 3\}$ and $B = \{1, 2, 4\}$, then $P(A) = 3/4$, $P(B) = 3/4$, and $P(A \cap B) = 1/2$. Hence $P(B|A) = 2/3$, which is less than $P(B)$.

A very interesting case occurs when knowledge that A occurs does not change the odds that B occurs. As an example of this let $A = \{1, 2\}$ and $B = \{1, 3\}$; then $P(A) = 1/2$, $P(B) = 1/2$, $P(A \cap B) = 1/4$, and therefore $P(B \mid A) = 1/2$. Events such as these, for which the conditional probability is the same as the unconditional probability, are said to be independent.

Let A and B now be any two events in a general probability space, and suppose that $P(A) \neq 0$. We can then define A and B to be independent if $P(B \mid A) = P(B)$. Since $P(B \mid A) = P(B \cap A)/P(A)$ we see that if A and B are independent then

$$P(A \cap B) = P(A)P(B). \tag{19}$$

Since (19) makes sense even if $P(A) = 0$ and is also symmetric in the letters A and B, it leads to a preferred definition of independence.

Definition 4 *Two events A and B are independent if and only if*

$$P(A \cap B) = P(A)P(B).$$

We can consider a similar problem for three sets A, B, and C. Take $\Omega = \{1, 2, 3, 4\}$ and assign probability 1/4 to each point. Let $A = \{1, 2\}$, $B = \{1, 3\}$, and $C = \{1, 4\}$. Then we leave it as an exercise to show that the pairs of events A and B, A and C, and B and C are independent. We say that the events A, B, and C are *pairwise* independent. On the other hand, $P(C) = 1/2$ and

$$P(C \mid A \cap B) = 1.$$

Thus a knowledge that the event $A \cap B$ occurs increases the odds that C occurs. In this sense the events A, B, and C fail to be *mutually* independent. In general, three events A, B, and C are mutually independent if they are pairwise independent and if

$$P(A \cap B \cap C) = P(A)P(B)P(C).$$

We leave it as an exercise to show that if A, B, and C are mutually independent and $P(A \cap B) \neq 0$, then $P(C \mid A \cap B) = P(C)$.

More generally we define $n \geq 3$ events $A_1, A_2, \ldots, A_n$ to be mutually independent if

$$P(A_1 \cap \cdots \cap A_n) = P(A_1) \cdots P(A_n)$$

and if any subcollection containing at least two but fewer than n events are mutually independent.

Example 8. Let S be the square $0 \leq x \leq 1$, $0 \leq y \leq 1$ in the plane. Consider the uniform probability space on the square, and let A be the event

$$\{(x, y): 0 \leq x \leq 1/2, 0 \leq y \leq 1\}$$

and B be the event

$$\{(x, y): 0 \leq x \leq 1, 0 \leq y \leq 1/4\}.$$

Show that A and B are independent events.

To show this, we compute $P(A)$, $P(B)$, and $P(A \cap B)$, and show that $P(A \cap B) = P(A)P(B)$. Now A is a subrectangle of the square S having area $1/2$ and B is a subrectangle of the square S having area $1/4$, so $P(A) = 1/2$ and $P(B) = 1/4$. Since

$$A \cap B = \{(x, y): 0 \leq x \leq 1/2, 0 \leq y \leq 1/4\}$$

is a subrectangle of the square S having area $1/8$, $P(A \cap B) = 1/8$ and we see that A and B are independent events as was asserted.

The notion of independence is frequently used to construct probability spaces corresponding to repetitions of the same experiment. This matter will be dealt with more fully in Chapter 3. We will be content here to examine the simplest situation, namely, experiments (such as tossing a possibly biased coin) that can result in only one of two possible outcomes—success or failure.

In an experiment such as tossing a coin n times, where success and failure at each toss occur with probabilities p and $1 - p$ respectively, we intuitively believe that the outcome of the ith toss should have no influence on the outcome of the other tosses. We now wish to construct a probability space corresponding to the compound experiment of an n-fold repetition of our simple given experiment that incorporates our intuitive beliefs.

Since each of the n trials can result in either success or failure, there is a total of 2^n possible outcomes to the compound experiment. These may be represented by an n-tuple $(x_1, \ldots, x_n)$, where $x_i = 1$ or 0 according as the ith trial yields a success or failure. We take the set Ω to be the collection of all such n-tuples. The σ-field $\mathscr{A}$ is taken to be all subsets of Ω.

We now come to the assignment of a probability measure. To do this it is only necessary to assign probabilities to the 2^n one-point sets $\{(x_1, \ldots, x_n)\}$. Suppose the n-tuple $(x_1, \ldots, x_n)$ is such that exactly k of the x_i's have the value 1; for simplicity, say $x_1 = x_2 = \cdots = x_k = 1$ and the other x_i's have the value 0. Then if A_i denotes the event that the ith trial, $1 \le i \le n$, is a success, we see that

$$\{(\underbrace{1, 1, \ldots, 1}_{k}, \underbrace{0, \ldots, 0}_{n-k})\} = A_1 \cap \cdots \cap A_k \cap A_{k+1}^c \cap \cdots \cap A_n^c.$$

According to our intuitive views, the events $A_1, \ldots, A_k, A_{k+1}^c, \ldots, A_n^c$ are to be mutually independent and $P(A_i) = p$, $1 \le i \le n$. Thus we should assign P so that

$$\begin{aligned} P(\{(1, 1, \ldots, 1, 0, \ldots, 0)\}) &= P(A_1) \cdots P(A_k) P(A_{k+1}^c) \cdots P(A_n^c) \\ &= p^k(1 - p)^{n-k}. \end{aligned}$$

By the same reasoning, we see that if the n-tuple $(x_1, \ldots, x_n)$ is such that exactly k of the x_i's have the value 1, then P should be such that

$$P(\{(x_1, \ldots, x_n)\}) = p^k(1 - p)^{n-k}.$$

Let us now compute the probability that exactly k of the n trials result in a success. Note carefully that this differs from the probability that k specified trials result in successes and the other $n - k$ trials result in failures. Let B_k denote the event that exactly k of the n trials are successes. Since every choice of a specified sequence having k successes has probability $p^k(1 - p)^{n-k}$, the event B_k has probability $P(B_k) = C(k, n)p^k(1 - p)^{n-k}$, where $C(k, n)$ is the number of sequences $(x_1, \ldots, x_n)$ in which exactly k of the x_i's have value 1. The computation of $C(k, n)$ is a simple combinatorial problem that will be solved in Section 2.4. There it will be shown that

$$C(k, n) = \frac{n!}{k!\,(n - k)!}, \qquad 0 \le k \le n. \tag{20}$$

Recall that $0! = 1$ and that, for any positive integer m,

$$m! = m(m - 1) \cdots 1.$$

The quantity $n!/k!(n - k)!$ is usually written as $\binom{n}{k}$ (the binomial coefficient). Thus

$$P(B_k) = \binom{n}{k} p^k(1 - p)^{n-k}. \tag{21}$$

Various applied problems are modeled by independent success–failure trials. Typical is the following.

Example 9. Suppose a machine produces bolts, 10% of which are defective. Find the probability that a box of 3 bolts contains at most one defective bolt.

To solve the problem we assume that the production of bolts constitutes repeated independent success–failure trials with a defective bolt being a success. The probability of success in this case is then .1. Let B_0 be the event that none of the three bolts are defective and let B_1 be the event that exactly one of the three bolts is defective. Then $B_0 \cup B_1$ is the event that at most one bolt is defective. Since the events B_0 and B_1 are clearly disjoint, it follows that

$$\begin{aligned} P(B_0 \cup B_1) &= P(B_0) + P(B_1) \\ &= \binom{3}{0}(.1)^0(.9)^3 + \binom{3}{1}(.1)^1(.9)^2 \\ &= (.9)^3 + 3(.1)(.9)^2 \\ &= .972. \end{aligned}$$

Exercises

1 Let $(\Omega, \mathscr{A}, P)$ be a probability space, where $\mathscr{A}$ is the σ-field of all subsets of Ω and P is a probability measure that assigns probability $p > 0$ to each one-point set of Ω.
   (a) Show that Ω must have a finite number of points. *Hint:* show that Ω can have no more than p^{-1} points.
   (b) Show that if n is the number of points in Ω then p must be n^{-1}.

2 A model for a random spinner can be made by taking a uniform probability space on the circumference of a circle of radius 1, so that the probability that the pointer of the spinner lands in an arc of length s is $s/2\pi$. Suppose the circle is divided into 37 zones numbered $1, 2, \ldots, 37$. Compute the probability that the spinner stops in an even zone.

3 Let a point be picked at random in the unit square. Compute the probability that it is in the triangle bounded by $x = 0$, $y = 0$, and $x + y = 1$.

4 Let a point be picked at random in the disk of radius 1. Find the probability that it lies in the angular sector from 0 to $\pi/4$ radians.

5 In Example 2 compute the following probabilities:
   (a) No disintegration occurs before time 10.
   (b) There is a disintegration before time 2 or a disintegration between times 3 and 5.

6 A box contains 10 balls, numbered 1 through 10. A ball is drawn from the box at random. Compute the probability that the number on the ball was either 3, 4, or 5.

7 Suppose two dice are rolled once and that the 36 possible outcomes are equally likely. Find the probability that the sum of the numbers on the two faces is even.

8 Suppose events A and B are such that $P(A) = 2/5$, $P(B) = 2/5$, and $P(A \cup B) = 1/2$. Find $P(A \cap B)$.

9 If $P(A) = 1/3$, $P(A \cup B) = 1/2$, and $P(A \cap B) = 1/4$, find $P(B)$.

10 Suppose a point is picked at random in the unit square. Let A be the event that it is in the triangle bounded by the lines $y = 0$, $x = 1$, and $x = y$, and B be the event that it is in the rectangle with vertices $(0, 0)$, $(1, 0)$, $(1, 1/2)$, $(0, 1/2)$. Compute $P(A \cup B)$ and $P(A \cap B)$.

11 A box has 10 balls numbered $1, 2, \ldots, 10$. A ball is picked at random and then a second ball is picked at random from the remaining nine balls. Find the probability that the numbers on the two selected balls differ by two or more.

12 If a point selected at random in the unit square is known to be in the triangle bounded by $x = 0$, $y = 0$, and $x + y = 1$, find the probability that it is also in the triangle bounded by $y = 0$, $x = 1$, and $x = y$.

13 Suppose we have four chests each having two drawers. Chests 1 and 2 have a gold coin in one drawer and a silver coin in the other drawer. Chest 3 has two gold coins and chest 4 has two silver coins. A chest is selected at random and a drawer opened. It is found to contain a gold coin. Find the probability that the other drawer has
(a) a silver coin;
(b) a gold coin.

14 A box has 10 balls, 6 of which are black and 4 of which are white. Three balls are removed from the box, their color unnoted. Find the probability that a fourth ball removed from the box is white. Assume that the 10 balls are equally likely to be drawn from the box.

15 With the same box composition as in Exercise 14, find the probability that all three of the removed balls will be black if it is known that at least one of the removed balls is black.

16 Suppose a factory has two machines A and B that make 60% and 40% of the total production, respectively. Of their output, machine A produces 3% defective items, while machine B produces 5% defective items. Find the probability that a given defective part was produced by machine B.

17 Show by induction on n that the probability of selecting a red ball at any trial n in Polya's scheme (Example 7) is $r(b + r)^{-1}$.

18 A student is taking a multiple choice exam in which each question has 5 possible answers, exactly one of which is correct. If the student knows the answer he selects the correct answer. Otherwise he selects one answer at random from the 5 possible answers. Suppose that the student knows the answer to 70% of the questions.
(a) What is the probability that on a given question the student gets the correct answer?

(b) If the student gets the correct answer to a question, what is the probability that he knows the answer?

19 Suppose a point is picked at random in the unit square. If it is known that the point is in the rectangle bounded by $y = 0$, $y = 1$, $x = 0$, and $x = 1/2$, what is the probability that the point is in the triangle bounded by $y = 0$, $x = 1/2$, and $x + y = 1$?

20 Suppose a box has r red and b black balls. A ball is chosen at random from the box and then a second ball is drawn at random from the remaining balls in the box. Find the probability that
(a) both balls are red;
(b) the first ball is red and the second is black;
(c) the first ball is black and the second is red;
(d) both balls are black.

21 A box has 10 red balls and 5 black balls. A ball is selected from the box. If the ball is red, it is returned to the box. If the ball is black, it and 2 additional black balls are added to the box. Find the probability that a second ball selected from the box is
(a) red; (b) black.

22 Two balls are drawn, with replacement of the first drawn ball, from a box containing 3 white and 2 black balls.
(a) Construct a sample space for this experiment with equally likely sample points.
(b) Calculate the probability that both balls drawn will be the same color.
(c) Calculate the probability that at least one of the balls drawn will be white.

23 Work Exercise 22 if the first ball is not replaced.

24 Work Exercise 22 by constructing a sample space based on 4 sample points corresponding to white and black for each drawing.

25 Box I contains 2 white balls and 2 black balls, box II contains 2 white balls and 1 black ball, and box III contains 1 white ball and 3 black balls.
(a) One ball is selected from each box. Calculate the probability of getting all white balls.
(b) One box is selected at random and one ball drawn from it. Calculate the probability that it will be white.
(c) In (b), calculate the probability that the first box was selected given that a white ball is drawn.

26 A box contains 3 white balls and 2 black balls. Two balls are drawn from it without replacement.
(a) Calculate the probability that the second ball is black given that the first ball is black.
(b) Calculate the probability that the second ball is the same color as the first ball.

(c) Calculate the probability that the first ball is white given that the second ball is white.

27 A college is composed of 70% men and 30% women. It is known that 40% of the men and 60% of the women smoke cigarettes. What is the probability that a student observed smoking a cigarette is a man?

28 Assume that cars are equally likely to be manufactured on Monday, Tuesday, Wednesday, Thursday, or Friday. Cars made on Monday have a 4% chance of being "lemons"; cars made on Tuesday, Wednesday or Thursday have a 1% chance of being lemons; and cars made on Friday have a 2% chance of being lemons. If you bought a car and it turned out to be a lemon, what is the probability it was manufactured on Monday?

29 Suppose there were a test for cancer with the property that 90% of those with cancer reacted positively whereas 5% of those without cancer react positively. Assume that 1% of the patients in a hospital have cancer. What is the probability that a patient selected at random who reacts positively to this test actually has cancer?

30 In the three chests problem discussed in Example 6, compute the probability that the second drawer has a silver coin given that the first drawer has a gold coin.

31 In Polya's urn scheme (Example 7) given that the second ball was red, find the probability that
(a) the first ball was red;
(b) the first ball was black.

32 Suppose three identical and perfectly balanced coins are tossed once. Let A_i be the event that the ith coin lands heads. Show that the events A_1, A_2, and A_3 are mutually independent.

33 Suppose the six faces of a die are equally likely to occur and that the successive die rolls are independent. Construct a probability space for the compound experiment of rolling the die three times.

34 Let A and B denote two independent events. Prove that A and B^c, A^c and B, and A^c and B^c are also independent.

35 Let $\Omega = \{1, 2, 3, 4\}$ and assume each point has probability 1/4. Set $A = \{1, 2\}$, $B = \{1, 3\}$, $C = \{1, 4\}$. Show that the pairs of events A and B, A and C, and B and C are independent.

36 Suppose A, B, and C are mutually independent events and $P(A \cap B) \neq 0$. Show that $P(C \mid A \cap B) = P(C)$.

37 Experience shows that 20% of the people reserving tables at a certain restaurant never show up. If the restaurant has 50 tables and takes 52 reservations, what is the probability that it will be able to accommodate everyone?

38 A circular target of unit radius is divided into four annular zones with outer radii 1/4, 1/2, 3/4, and 1, respectively. Suppose 10 shots are fired independently and at random into the target.

(a) Compute the probability that at most three shots land in the zone bounded by the circles of radius 1/2 and radius 1.
(b) If 5 shots land inside the disk of radius 1/2, find the probability that at least one is in the disk of radius 1/4.

39 A machine consists of 4 components linked in parallel, so that the machine fails only if all four components fail. Assume component failures are independent of each other. If the components have probabilities .1, .2, .3, and .4 of failing when the machine is turned on, what is the probability that the machine will function when turned on?

40 A certain component in a rocket engine fails 5% of the time when the engine is fired. To achieve greater reliability in the engine working, this component is duplicated n times. The engine then fails only if all of these n components fail. Assume the component failures are independent of each other. What is the smallest value of n that can be used to guarantee that the engine works 99% of the time?

41 A symmetric die is rolled 3 times. If it is known that face 1 appeared at least once what is the probability that it appeared exactly once?

42 In a deck of 52 cards there are 4 kings. A card is drawn at random from the deck and its face value noted; then the card is returned. This procedure is followed 4 times. Compute the probability that there are exactly 2 kings in the 4 selected cards if it is known that there is at least one king in those selected.

43 Show that if A, B, and C are three events such that $P(A \cap B \cap C) \neq 0$ and $P(C \mid A \cap B) = P(C \mid B)$, then $P(A \mid B \cap C) = P(A \mid B)$.

44 A man fires 12 shots independently at a target. What is the probability that he hits the target at least once if he has probability 9/10 of hitting the target on any given shot?

45 A die is rolled 12 times. Compute the probability of getting
(a) 2 sixes;
(b) at most two sixes.

46 Suppose the probability of hitting a target is 1/4. If eight shots are fired at the target, what is the probability that the target is hit at least twice?

47 In Exercise 44, what is the probability that the target is hit at least twice if it is known that it is hit at least once?

2 Combinatorial Analysis

Recall from Section 1.2 that a symmetric probability space having s points is the model used for choosing a point at random from a set S having s points. Henceforth when we speak of choosing a point at random from a finite set S, we shall mean that the probability assigned to each one-point set is s^{-1}, and hence the probability assigned to a set A having j points is j/s.

Let $N(A)$ denote the number of points in A. Since $P(A) = N(A)/s$, the problem of computing $P(A)$ is equivalent to that of computing $N(A)$. The procedure for finding $P(A)$ is to "count" the number of points in A and divide by the total number of points s. However, sometimes the procedure is reversed. If by some means we know $P(A)$, then we can find $N(A)$ by the formula $N(A) = sP(A)$. This reverse procedure will be used several times in the sequel.

The computation of $N(A)$ is easy if A has only a few points, for in that case we can just enumerate all the points in A. But even if A has only a moderate number of points, the method of direct enumeration becomes intractable, and so some simple rules for counting are desirable. Our purpose in this chapter is to present a nontechnical systematic discussion of techniques that are elementary and of wide applicability. This subject tends to become difficult quite rapidly, so we shall limit our treatment to those parts of most use in probability theory. The first four sections in this chapter contain the essential material, while the last four sections contain optional and somewhat more difficult material.

2.1. Ordered samples

Suppose we have two sets S and T. If S has m distinct points $s_1, s_2, \ldots, s_m$ and T has n distinct points $t_1, t_2, \ldots, t_n$, then the number of pairs (s_i, t_j) that can be formed by taking one point from the set S and a second from the set T is mn. This is clear since any given element of the set S can be associated with any of the n elements from the set T.

Example 1. If $S = \{1, 2\}$ and $T = \{1, 2, 3\}$, then there are six pairs: $(1, 1)$, $(1, 2)$, $(1, 3)$, $(2, 1)$, $(2, 2)$, $(2, 3)$. Note carefully that the pair $(1, 2)$ is distinct from the pair $(2, 1)$.

More generally, suppose we have n sets $S_1, S_2, \ldots, S_n$ having $s_1, s_2, \ldots, s_n$ distinct points, respectively. Then the number of n-tuples $(x_1, x_2, \ldots, x_n)$ that can be formed where x_1 is an element from S_1, x_2 an element from $S_2, \ldots,$ and x_n an element from S_n is $s_1 s_2 \cdots s_n$. This is a quite obvious extension of the case for $n = 2$ discussed above. (A formal proof that the number of n-tuples is $s_1 s_2 \cdots s_n$ could be carried out by induction on n.)

An important special case occurs when each of the sets S_i, $1 \leq i \leq n$, is the same set S having s distinct points. There are then s^n n-tuples $(x_1, x_2, \ldots, x_n)$ where each x_i is one of the points of S.

Example 2. $S = \{1, 2\}$ and $n = 3$. Then there are eight n-tuples: (1, 1, 1), (1, 1, 2), (1, 2, 1), (1, 2, 2), (2, 1, 1), (2, 1, 2), (2, 2, 1), (2, 2, 2).

The special case when the sets S_i, $1 \leq i \leq n$, are the same set can be approached from a different point of view. Suppose a box has s distinct balls labeled $1, 2, \ldots, s$. A ball is drawn from the box, its number noted and the ball is returned to the box. The procedure is repeated n times. Each of the n draws yields a number from 1 to s. The outcome of the n draws can be recorded as an n-tuple $(x_1, x_2, \ldots, x_n)$, where x_1 is the number on the 1st ball drawn, x_2 that on the 2nd, etc. In all, there are s^n possible n-tuples. This procedure is called *sampling with replacement* from a population of s distinct objects. The outcome $(x_1, x_2, \ldots, x_n)$ is called a sample of size n drawn from a population of s objects with replacement. We speak of *random sampling with replacement* if we assume that all of the s^n possible samples possess the same probability or, in traditional language, are equally likely to occur.

Example 3. A perfectly balanced coin is tossed n times. Find the probability that there is at least one head.

Presumably the statement that the coin is perfectly balanced implies that the probability of getting a head on a given toss is 1/2. If this is so, and if we assume that flipping the coin n times is equivalent to drawing a random sample of size n from a population of the two objects {H, T}, then each of the 2^n possible outcomes is equally likely. Let A be the event that there is at least one head, and let A_i be the event that the ith toss results in a head. Then $A = \bigcup_{i=1}^n A_i$. But

$$\begin{aligned} P(A) &= 1 - P(A^c) \\ &= 1 - P\left(\left(\bigcup_{i=1}^{n} A_i\right)^c\right) \\ &= 1 - P\left(\bigcap_{i=1}^{n} A_i^c\right) \end{aligned}$$

and $\bigcap_{i=1}^n A_i^c$ occurs if and only if all of the n tosses yield tails. Thus $P(\bigcap_{i=1}^n A_i^c) = 2^{-n}$, so $P(A) = 1 - 2^{-n}$.

Let S denote a set having s distinct objects. We select an object from S and note which object it is, but now suppose we do not return it to the set. If we repeat this procedure we will then make a selection from the remaining $(s - 1)$ objects. Suppose the procedure is repeated $n - 1$ additional times, so that altogether n objects are selected. (Obviously we must have $n \leq s$ in this case.) Once again we may record the outcome as an n-tuple $(x_1, x_2, \ldots, x_n)$, but this time the numbers $x_1, x_2, \ldots, x_n$ must be distinct; there can be no duplications in our sample. The first object selected can be any one of s objects, the second object can be any one of the remaining $s - 1$ objects, the third can be any one of the remaining $s - 2$ objects, etc., so in all there are $(s)_n = s(s - 1) \cdots (s - n + 1)$ different possible outcomes to the experiment. This procedure is referred to as *sampling without replacement* n times from a population of s distinct objects. We speak of a *random sample of size n drawn from a population of s objects without replacement* if we assume that each of these $(s)_n$ outcomes is equally likely.

We have denoted the product $s(s - 1) \cdots (s - n + 1)$ by the symbol $(s)_n$. In particular, $(s)_s = s(s - 1) \cdots 1 = s!$ Now drawing a sample of size s from a population of s distinct objects is equivalent to writing down the numbers, $1, 2, \ldots, s$ in some order. Thus $s!$ represents the number of different orderings (or permutations) of s objects.

Suppose a random sample of size n is chosen from a set of s objects with replacement. We seek the probability of the event A that in the sample no point appears twice. The problem is easily solved. The number of samples of size n with replacement is s^n. Of these s^n random samples the number in which no point appears twice is the same as the number of samples of size n drawn from s objects *without replacement*, i.e., $(s)_n$. Thus since all the s^n samples are equally likely, we find that the required probability is

$$\frac{(s)_n}{s^n} = \frac{s(s-1) \cdots (s - n + 1)}{s^n} \tag{1}$$

$$= \left(1 - \frac{1}{s}\right)\left(1 - \frac{2}{s}\right) \cdots \left(1 - \frac{n-1}{s}\right).$$

Example 4. A novel and rather surprising application of (1) is the so-called birthday problem. Assume that people's birthdays are equally likely to occur among the 365 days of the year. (Here we ignore leap years and the fact that birth rates are not exactly uniform over the year.) Find the probability p that no two people in a group of n people will have a common birthday.

In this problem $s = 365$, so by applying (1) we see that

$$p = \left(1 - \frac{1}{365}\right)\left(1 - \frac{2}{365}\right) \cdots \left(1 - \frac{n-1}{365}\right).$$

The numerical consequences are quite unexpected. Even for n as small as 23, $p < 1/2$, and for $n = 56$, $p = .01$. That is, in a group of 23 people the probability that at least two people have a common birthday exceeds 1/2. In a group of 56 people, it is almost certain that two have the same birthday.

If we have a population of s objects, there are s^n samples of size n that can be drawn with replacement and $(s)_n$ samples of size n that can be drawn without replacement. If s is large compared to n, there is little difference between random sampling by these two methods. Indeed, we see from (1) that for any fixed n,

$$\lim_{s \to \infty} \frac{(s)_n}{s^n} = \lim_{s \to \infty} \left(1 - \frac{1}{s}\right) \cdots \left(1 - \frac{n-1}{s}\right) = 1. \tag{2}$$

(For more precise estimates see Exercise 12.)

2.2. Permutations

Suppose we have n distinct boxes and n distinct balls. The total number of ways of distributing the n balls into the n boxes in such a manner that *each box has exactly one ball* is $n!$. To say that these n balls are distributed at random into the n boxes with one ball per box means that we assign probability $1/n!$ to each of these possible ways. Suppose this is the case. What is the probability that a specified ball, say ball i, is in a specified box, say box j? If ball i is in box j, this leaves $(n - 1)$ boxes and $(n - 1)$ balls to be distributed into them so that exactly one ball is in each box. This can be done in $(n - 1)!$ ways, so the required probability is $(n - 1)!/n! = 1/n$.

Another way of looking at this result is as follows. If we have n distinct objects and we randomly permute them among themselves, then the probability that a specified object is in a specified position has probability $1/n$. Indeed, here the positions can be identified with the boxes and the objects with the balls.

The above considerations are easily extended from 1 to $k \geq 1$ objects. If n objects are randomly permuted among themselves, the probability that k specified objects are in k specified positions is $(n - k)!/n!$. We leave the proof of this fact to the reader.

Problems involving random permutations take on a variety of forms when stated as word problems. Here are two examples:

(a) A deck of cards labeled $1, 2, \ldots, n$ is shuffled, and the cards are then dealt out one at a time. What is the probability that for some specified i, the ith card dealt is the card labeled i?

(b) Suppose 10 couples arrive at a party. The boys and girls are then paired off at random. What is the probability that exactly k specified boys end up with their own girls?

A more sophisticated problem involving random permutations is to find the probability that there are exactly k "matches." To use our usual picturesque example of distributing balls in boxes, the problem is to find the probability that ball i is in box i for exactly k different values of i.

The problem of matchings can be solved in a variety of ways. We postpone discussion of this problem until Section 2.6.

2.3. Combinations (unordered samples)

A poker hand consists of five cards drawn from a deck of 52 cards. From the point of view of the previous discussion there would be $(52)_5$ such hands. However, in arriving at this count different orderings of the same five cards are considered different hands. That is, the hand 2, 3, 4, 5, 6 of spades in that order is considered different from the hand 2, 4, 3, 5, 6 of spades in that order. From the point of view of the card game, these hands are the same. In fact all of the 5! permutations of the same five cards are equivalent. Of the $(52)_5$ possible hands, exactly 5! of them are just permutations of these same five cards. Similarly, for any given set of five cards there are 5! different permutations. Thus the total number of poker hands, *disregarding the order in which the cards appear*, is $(52)_5/5!$. In this new count two hands are considered different if and only if they differ as sets of objects, i.e., they have at least one element different. For example, among the $(52)_5/5!$ poker hands, the hands (2, 3, 4, 5, 6) of spades and (3, 2, 4, 5, 6) of spades are the same, but the hands (2, 3, 4, 5, 7) of spades and (2, 3, 4, 5, 6) of spades are different.

More generally, suppose we have a set S of s distinct objects. Then, as previously explained, there are $(s)_r$ distinct samples of size r that can be drawn from S without replacement. Each distinct subset $\{x_1, \ldots, x_r\}$ of r objects from S can be ordered (rearranged) in $r!$ different ways. If we choose to ignore the order that the objects appear in the sample, then these $r!$ reorderings of $x_1, \ldots, x_r$ would be considered the same. There are therefore $(s)_r/r!$ *different samples of size r that can be drawn without replacement and without regard to order from a set of s distinct objects.*

The quantity $(s)_r/r!$ is usually written by means of the binomial coefficient symbol

$$\frac{(s)_r}{r!} = \binom{s}{r}.$$

A-HA

Observe that for $r = 0, 1, 2, \ldots, s$

$$\binom{s}{r} = \frac{(s)_r}{r!} = \frac{s!}{r!(s-r)!}.$$

We point out here for future use that $\binom{a}{r}$ is well defined for any real number a and nonnegative integer r by

$$\binom{a}{r} = \frac{(a)_r}{r!} = \frac{a(a-1)\cdots(a-r+1)}{r!}, \tag{3}$$

where $0!$ and $(a)_0$ are both defined to be 1.

Example 5.

$$\binom{-\pi}{3} = \frac{(-\pi)(-\pi-1)(-\pi-2)}{3!}$$

$$= -\frac{\pi(\pi+1)(\pi+2)}{3!}.$$

Observe that if a is a positive integer then $\binom{a}{r} = 0$ if $r > a$. We adopt the convention that $\binom{a}{r} = 0$ if r is a negative integer. Then $\binom{a}{r}$ is defined for all real numbers a and all integers r.

As previously observed, when s is a positive integer and r is a nonnegative integer, it is useful to think of $\binom{s}{r}$ as the number of ways we can draw a sample of size r from a population of s distinct elements without replacement and without regard to the order in which these r objects were chosen.

Example 6. Consider the set of numbers $\{1, 2, \ldots, n\}$. Then if $1 \le r \le n$, there are exactly $\binom{n}{r}$ choices of numbers $i_1, i_2, \ldots, i_r$ such that $1 \le i_1 < i_2 < \cdots < i_r \le n$. Indeed, each of the $(n)_r$ choices of r *distinct* numbers from 1 to n has $r!$ reorderings exactly one of which satisfies the requirement. Thus the number of distinct choices of numbers satisfying the requirement is the same as the number of distinct subsets of size r that can be drawn from the set $\{1, 2, \ldots, n\}$.

Example 7. Committee membership. The mathematics department consists of 25 full professors, 15 associate professors, and 35 assistant professors. A committee of 6 is selected at random from the faculty of the department. Find the probability that all the members of the committee are assistant professors.

In all, there are 75 faculty members. The committee of 6 can be chosen from the 75 in $\binom{75}{6}$ ways. There are 35 assistant professors, and the 6 that are on the committee can be chosen from the 35 in $\binom{35}{6}$ ways. Thus the required probability is $\binom{35}{6}\Big/\binom{75}{6}$. Calculations yield the approximate value of .01; therefore the tenure staff (associate and full professors) need not worry unduly about having no representation.

Example 8. Consider a poker hand of five cards. Find the probability of getting four of a kind (i.e., four cards of the same face value) assuming the five cards are chosen at random.

We may solve the problem as follows.

There are $\binom{52}{5}$ different hands, which are to be equally likely. Thus Ω will have $\binom{52}{5}$ points. For the desired event to occur we must have four cards of the same face value. There are 13 different choices for the value that the four of a kind is to have, namely 2, 3, 4, 5, 6, 7, 8, 9, 10, J, Q, K, A. For each such choice (which determines four of the five cards in the desired hand) there are 48 other cards from which to choose the 5th card of the hand. Since any of the 13 choices of the four of a kind can be paired with any of the 48 choices remaining for the 5th card, in all there are (13)(48) possible ways of getting a poker hand with four of the five cards equal. The desired probability is therefore

$$\frac{(13)(48)}{\binom{52}{5}} \approx 2.40 \times 10^{-4}.$$

Example 9. Suppose n balls are distributed into n boxes so that all of the n^n possible arrangements are equally likely. Compute the probability that only box 1 is empty.

The probability space in this case consists of n^n equally likely points. Let A be the event that only box 1 is empty. This can happen only if the n balls are in the remaining $n - 1$ boxes in such a manner that no box is empty. Thus, exactly one of these $(n - 1)$ boxes must have two balls, and the remaining $(n - 2)$ boxes must have exactly one ball each. Let B_j be the event that box j, $j = 2, 3, \ldots, n$, has two balls, box 1 has no balls, and the remaining $(n - 2)$ boxes have exactly one ball each. Then the B_j are disjoint and $A = \bigcup_{j=2}^{n} B_j$. To compute $P(B_j)$ observe that the two balls put in box j can be chosen from the n balls in $\binom{n}{2}$ ways. The $(n - 2)$ balls in the remaining $(n - 2)$ boxes can be rearranged in $(n - 2)!$ ways.

Thus the number of distinct ways we can put two balls into box j, no ball in box 1, and exactly one ball in each of the remaining boxes is $\binom{n}{2}(n-2)!$. Hence

$$P(B_j) = \frac{\binom{n}{2}(n-2)!}{n^n}$$

and consequently

$$P(A) = \frac{(n-1)\binom{n}{2}(n-2)!}{n^n} = \frac{\binom{n}{2}(n-1)!}{n^n}.$$

2.4. Partitions

A large variety of combinatorial problems involving unordered samples are of the following type. A box has r red balls and b black balls. A random sample of size n is drawn from the box without replacement. What is the probability that this sample contains exactly k red balls (and hence $n - k$ black balls)?

To solve the problem we argue as follows. We are interested only in the total number of red balls and black balls in the sample and not in the order in which these balls were drawn. That is, we are dealing with sampling without replacement and without regard to order. We can, therefore, take our probability space for this problem to be the collection of all $\binom{b+r}{n}$ samples of size n that can be drawn in this manner from the $b + r$ balls in the population. Each of these $\binom{b+r}{n}$ samples is assigned the same probability $\binom{b+r}{n}^{-1}$. We must now compute the number of ways in which a sample of size n can be drawn so as to have exactly k red balls. The k red balls can be chosen from the r red balls in $\binom{r}{k}$ ways without regard to order, and the $n - k$ black balls can be chosen from the b black balls without regard to order in $\binom{b}{n-k}$ ways. Since each choice of k red balls could be paired with each choice of $n - k$ black balls there are, therefore, a total of $\binom{r}{k}\binom{b}{n-k}$ possible choices. Thus the desired probability is

$$\frac{\binom{r}{k}\binom{b}{n-k}}{\binom{r+b}{n}}.$$

The essence of this type of problem is that the population (in this case the balls) is partitioned into two classes (red and black balls). A random sample of a certain size is taken and we require the probability that the sample will contain a specified number of items in each of the two classes. In some problems of this type the two classes are not explicitly specified, but they can be recognized when the language of the problem is analyzed.

Example 10. A poker hand has five cards drawn from an ordinary deck of 52 cards. Find the probability that the poker hand has exactly 2 kings.

To solve the problem note that there are $\binom{52}{5}$ poker hands. In the deck there are 4 kings and 48 other cards. This partitions the cards into two classes, kings and non-kings, having respectively 4 and 48 objects each. The poker hand is a sample of size 5 drawn without replacement and without order from the 52 cards. The problem thus is to find the probability that the sample has 2 members of the first class and 3 members of the second class. Hence the required probability is

$$\frac{\binom{4}{2}\binom{48}{3}}{\binom{52}{5}} \approx 3.99 \times 10^{-2}.$$

Example 11. A deck of playing cards has 4 suits of 13 cards each, namely clubs, diamonds, hearts, and spades.

(a) What is the probability that in a hand of 5 cards exactly 3 are clubs?

(b) What is the probability that in a hand of 5 cards exactly 3 are of the same suit?

To solve problem (a) we note that the conditions of the problem partition the deck of 52 cards into 2 classes. Class one is the "clubs" having 13 members, and class two is "other than clubs" having 39 members. The 5 cards constitute a sample of size 5 from the population of 52 cards, and the problem demands that 3 of the 5 be from class one. Thus the required probability is

$$p = \frac{\binom{13}{3}\binom{39}{2}}{\binom{52}{5}} \approx 8.15 \times 10^{-2}.$$

To solve (b) let A_1 be the event that exactly 3 cards are clubs, A_2 the event that exactly 3 cards are diamonds, A_3 the event that exactly 3 cards are hearts, and A_4 the event that exactly 3 cards are spades. Then since there are only 5 cards in the hand, the events A_1, A_2, A_3, A_4 are mutually disjoint. Their union, $A_1 \cup A_2 \cup A_3 \cup A_4$, is just the event that of the 5 cards exactly 3 are of the same suit. Thus the required probability is $4p$.

Example 12. Consider again a poker hand of 5 cards. What is the probability that it is a full house (i.e., one pair of cards with equal face value and one triple of cards with equal face value), assuming that the cards are drawn at random from the deck?

To solve the problem we again note that there are $\binom{52}{5}$ poker hands each of which is equally likely. Of these we must now compute the number of ways in which we can have one pair and one triple. Consider the number of ways we can choose a *particular triple*, say 3 aces, and a *particular pair*, say 2 kings. The triple has 3 cards that are to be chosen without regard to order from the four aces and this can be done in $\binom{4}{3}$ ways. The pair has two cards to be drawn without regard to order from the four kings. This can be done in $\binom{4}{2}$ ways. The total number of ways then of drawing a hand having a triple of aces and a pair of kings is $\binom{4}{3}\binom{4}{2}$. Thus the probability of getting a poker hand that has a triple of aces and a pair of kings is $\binom{4}{3}\binom{4}{2}\Big/\binom{52}{5} = p$. Of course, this probability would be the same for any specified pair and any specified triple. Now the face value of the cards on the triple can be any of the possible 13, and the face value of the cards in the pair can be any of the 12 remaining possible face values. Since each of the 13 values for the triple can be associated with each of the 12 values for the pair, there are (13)(12) such choices. Each of these choices constitutes a disjoint event having probability p, so the required probability is

$$(13)(12)p = \frac{(13)(12)(4)(6)}{\binom{52}{5}} \approx 1.44 \times 10^{-3}.$$

Example 13. In a poker hand what is the probability of getting exactly two pairs? Here, a hand such as $(2, 2, 2, 2, x)$ does not count as two pairs but as a 4-of-a-kind.

To solve the problem we note that if the hand has two pairs, then two of the cards have the same face value x, two of the cards have the same face value $y \neq x$, and the fifth card has a different face value from x or y. Now there are 13 different face values. The face values of the two pairs can be chosen from them in $\binom{13}{2}$ ways. The other card can be any one of 11 face values. The two cards of value x can be chosen from the four of that value in $\binom{4}{2}$ ways and likewise for the 2 of value y. The remaining card of value z can be chosen from the four of that value in $\binom{4}{1} = 4$ ways. Thus the number of choices is $\binom{13}{2}(11)\binom{4}{2}\binom{4}{2}(4)$ so the desired probability is

$$\frac{\binom{13}{2}(11)(6)(6)(4)}{\binom{52}{5}} \approx 4.75 \times 10^{-2}.$$

In some problems involving partitioning, the classes are imagined as in the following.

Example 14. Suppose we have a box containing r balls numbered $1, 2, \ldots, r$. A random sample of size n is drawn without replacement and the numbers on the balls noted. These balls are then returned to the box, and a second random sample of size m is then drawn without replacement. Find the probability that the two samples had exactly k balls in common.

To solve this problem we can argue as follows. The effect of the first sample is to partition the balls into two classes, viz., those n selected and those $r - n$ not selected. (We can imagine that the n balls selected in the first sample are painted red before being tossed back). The problem is then of finding the probability that the sample of size m contains exactly k balls from the first class, so the desired probability is

$$\frac{\binom{n}{k}\binom{r-n}{m-k}}{\binom{r}{m}}.$$

If the argument were done in reverse, and we thought of the second sample as marking the balls, then we would find that the probability is

$$\frac{\binom{m}{k}\binom{r-m}{n-k}}{\binom{r}{n}}.$$

We leave it as an exercise to show that these two are equal.

We can easily extend our consideration of partitioning a population into two classes to partitioning it into $m \geq 2$ classes. Suppose we have a set of r objects such that each object is one of m possible types. The population consists of r_1 objects of type 1, r_2 objects of type $2, \ldots, r_m$ objects of type m, where $r_1 + r_2 + \cdots + r_m = r$. If a random sample of size n is drawn without replacement from the population of these r objects, what is the probability that the sample contains exactly k_1 objects of type $1, \ldots, k_m$ objects of type m, where $k_1 + \cdots + k_m = n$?

Once again the probability space is the collection of all $\binom{r}{n}$ equally likely samples of size n that can be drawn without replacement and without regard to order from the r objects in the population. The k_i objects of type i in the sample can be chosen from the r_i objects of that type without regard to order in $\binom{r_i}{k_i}$ ways. Thus the probability of choosing the sample with the specified composition is

$$\frac{\binom{r_1}{k_1}\binom{r_2}{k_2}\cdots\binom{r_m}{k_m}}{\binom{r}{n}}.$$

Example 15. In a hand of 13 cards chosen from an ordinary deck, find the probability that it is composed of exactly 3 clubs, 4 diamonds, 4 hearts, and 2 spades.

In this problem $r = 52$, $n = 13$. Let class 1 be clubs, class 2 diamonds, class 3 hearts, and class 4 spades. Then $m = 4$, $k_1 = 3$, $k_2 = 4$, $k_3 = 4$, and $k_4 = 2$, so the desired probability is

$$\frac{\binom{13}{3}\binom{13}{4}\binom{13}{4}\binom{13}{2}}{\binom{52}{13}}.$$

Example 16. Committee problem. In the committee problem discussed earlier, find the probability that the committee of 6 is composed of 2 full professors, 3 associate professors, and 1 assistant professor.

Using the same method as above, we find the answer to be

$$\frac{\binom{25}{2}\binom{15}{3}\binom{35}{1}}{\binom{75}{6}}.$$

2.5. Union of events*

Consider again the random permutation of n distinct objects. We say a match occurs at the ith position if the ith object is in the ith position. Let A_i be the event that there is a match at position i. Then $A = \bigcup_{i=1}^{n} A_i$ is the event that there is at least one match. We can compute $P(\bigcup_{i=1}^n A_i)$ for $n = 2$ by Equation (10) of Chapter 1 which states that

$$P(A_1 \cup A_2) = P(A_1) + P(A_2) - P(A_1 \cap A_2).$$

It is possible to use this formula to find a similar formula for $n = 3$. Let A_1, A_2, and A_3 be three events and set $B = A_1 \cup A_2$. Then

$$P(A_1 \cup A_2 \cup A_3) = P(B \cup A_3) = P(B) + P(A_3) - P(B \cap A_3).$$

Now

$$P(B) = P(A_1 \cup A_2) = P(A_1) + P(A_2) - P(A_1 \cap A_2). \tag{4}$$

Since $B \cap A_3 = (A_1 \cup A_2) \cap A_3 = (A_1 \cap A_3) \cup (A_2 \cap A_3)$, it follows that

$$P(B \cap A_3) = P(A_1 \cap A_3) + P(A_2 \cap A_3) - P(A_1 \cap A_2 \cap A_3). \tag{5}$$

Substituting (4) and (5) into the expression for $P(A_1 \cup A_2 \cup A_3)$, we see that

$$\begin{aligned} P(A_1 \cup A_2 \cup A_3) &= [P(A_1) + P(A_2) - P(A_1 \cap A_2)] + P(A_3) \\ &\quad - [P(A_1 \cap A_3) + P(A_2 \cap A_3) - P(A_1 \cap A_2 \cap A_3)] \\ &= [P(A_1) + P(A_2) + P(A_3)] \\ &\quad - [P(A_1 \cap A_2) + P(A_1 \cap A_3) + P(A_2 \cap A_3)] \\ &\quad + P(A_1 \cap A_2 \cap A_3). \end{aligned}$$

In order to express this formula more conveniently, we set

$$S_1 = P(A_1) + P(A_2) + P(A_3),$$

$$S_2 = P(A_1 \cap A_2) + P(A_1 \cap A_3) + P(A_2 \cap A_3),$$

and

$$S_3 = P(A_1 \cap A_2 \cap A_3).$$

Then

$$P(A_1 \cup A_2 \cup A_3) = S_1 - S_2 + S_3. \tag{6}$$

There is a generalization of (6) that is valid for all positive integers n. Let $A_1, \ldots, A_n$ be events. Define n numbers S_r, $1 \le r \le n$, by

$$S_r = \sum_{1 \le i_1 < \cdots < i_r \le n} P(A_{i_1} \cap \cdots \cap A_{i_r}).$$

Then in particular

$$S_1 = P(A_1) + \cdots + P(A_n),$$

$$S_2 = \sum_{i=1}^{n-1} \sum_{j=i+1}^{n} P(A_i \cap A_j),$$

and

$$S_n = P(A_1 \cap \cdots \cap A_n).$$

The desired formula for $P(\bigcup_{i=1}^n A_i)$ is given by:

$$P\left(\bigcup_{i=1}^{n} A_i\right) = \sum_{r=1}^{n} (-1)^{r-1} S_r \tag{7}$$

$$= S_1 - S_2 + \cdots + (-1)^{n-1} S_n.$$

The reader can easily check that this formula agrees with (6) if $n = 3$ and with Equation (10) of Chapter 1 if $n = 2$. The proof of (7) proceeds by induction, but is otherwise similar to that of (6). We will omit the details of the proof.

The sum S_1 has n terms, the sum S_2 has $\binom{n}{2}$ terms, and in general the sum S_r has $\binom{n}{r}$ terms. To see this, note that the rth sum is just the sum of the numbers $P(A_{i_1} \cap \cdots \cap A_{i_r})$ over all the values of the indices $i_1, i_2, \ldots, i_r$ such that $i_1 < i_2 < \cdots < i_r$. The indices take values between 1 and n. Thus the number of different values that these indices can take is the same as the number of ways we can draw r distinct numbers from n numbers without replacement and without regard to order.

2.6. Matching problems*

We now may easily solve the problem of the number of matches. Let A_i denote the event that a match occurs at the ith position and let p_n denote the probability that there are no matches. To compute $1 - p_n = P(\bigcup_{i=1}^n A_i)$, we need to compute $P(A_{i_1} \cap A_{i_2} \cap \cdots \cap A_{i_r})$ where $i_1, i_2, \ldots, i_r$ are r distinct numbers from $\{1, 2, \ldots, n\}$. But this probability is just the probability of a match at positions $i_1, i_2, \ldots, i_r$, and we have already found that the probability of this happening is $(n - r)!/n!$. Since the rth sum S_r has exactly $\binom{n}{r}$ terms we see that

$$P(A_1 \cup \cdots \cup A_n) = \sum_{r=1}^{n} \binom{n}{r} \frac{(n-r)!}{n!} (-1)^{r-1}$$

$$= \sum_{r=1}^{n} (-1)^{r-1} \frac{n!}{r!(n-r)!} \frac{(n-r)!}{n!}$$

$$= \sum_{r=1}^{n} \frac{(-1)^{r-1}}{r!};$$

that is,

$$(1 - p_n) = 1 - \frac{1}{2!} + \frac{1}{3!} - \cdots + \frac{(-1)^{n-1}}{n!}. \tag{8}$$

Using (8) we see that the probability, p_n, that there are no matches is

$$(9) \qquad p_n = 1 - 1 + \frac{1}{2!} - \frac{1}{3!} + \cdots + \frac{(-1)^n}{n!} = \sum_{k=0}^{n} \frac{(-1)^k}{k!}.$$

Now the right-hand side of (9) is just the first $n + 1$ terms of the Taylor expansion of e^{-1}. Therefore, we can approximate p_n by e^{-1} and get $1 - e^{-1} = .6321\ldots$ as an approximation to $(1 - p_n)$. It turns out that this approximation is *remarkably good* even for small values of n. In the table below we compute the values of $(1 - p_n)$ for various values of n.

n	3	4	5	6
$1 - p_n$	.6667	.6250	.6333	.6320

We thus have the remarkable result that the probability of at least one match among n randomly permuted objects is practically independent of n.

The problem of matches can be recast into a variety of different forms. One of the most famous of these is the following.

Two equivalent decks of cards are well shuffled and matched against each other. What is the probability of at least one match?

To solve the problem we need only observe that the first deck can be used to determine positions (boxes). With no loss of generality then we can assume the cards in the first deck are arranged in the order $1, 2, \ldots, n$. The cards in the second deck (the balls) are then matched against the positions determined by the first deck. A match occurs at position i if and only if the ith card drawn from the second deck is card number i.

Now that we know how to compute the probability p_n of no matches, we can easily find the probability $\beta_n(r)$ that there are exactly r matches. To solve the problem we first compute the probability that there are exactly r matches and that these occur at the first r places. This can happen only if there are no matches in the remaining $(n - r)$ places. The probability that there are no matches among j randomly permuted objects is p_j. Hence $j!\,p_j$ is the number of ways that j objects can be permuted among themselves so that there are no matches. (Why?) Since there is only one way of having r matches at the first r positions, the number of ways we can have exactly r matches at the first r positions and no matches at the remaining $(n - r)$ positions is $(n - r)!\,p_{n-r}$. Thus the required probability is

$$\alpha_r = \frac{(n - r)!}{n!}\,p_{n-r}.$$

The probability that there are exactly r matches and that these occur at any specified r positions is the same for all specifications, namely, α_r.

To solve the problem that there are exactly r matches, all that is now necessary is to realize that the events "exactly r matches occurring at positions $i_1, i_2, \ldots, i_r$," are disjoint events for the various choices of $i_1, i_2, \ldots, i_r$. The number of such choices is $\binom{n}{r}$. Thus the required probability is $\binom{n}{r}\alpha_r$. Hence, if $\beta_n(r)$ is the probability of exactly r matches among n randomly permuted objects, we find that

$$\beta_n(r) = \frac{n!\,\alpha_r}{r!\,(n-r)!} \tag{10}$$

$$= \frac{n!}{r!\,(n-r)!}\,\frac{(n-r)!\,p_{n-r}}{n!}$$

$$= \frac{p_{n-r}}{r!}$$

$$= \frac{1}{r!}\left[1 - 1 + \frac{1}{2!} + \cdots + \frac{(-1)^{n-r}}{(n-r)!}\right].$$

Using the approximation that p_{n-r} is approximately e^{-1} (which is very good even for $n - r$ moderately large) we find that

$$\beta_n(r) \approx \frac{e^{-1}}{r!}. \tag{11}$$

As a final illustration of these ideas, we compute the probability that there is a match in the jth place given that there are exactly r matches.

To solve this problem let A_j be the event that a match occurs at the jth place and let B_r be the event that there are exactly r matches. The desired probability is $P(A_j \mid B_r)$. From (10), $P(B_r) = p_{n-r}/r!$, so we need to compute $P(A_j \cap B_r)$. Now the event $A_j \cap B_r$ occurs if and only if there is a match in the jth place and exactly $(r-1)$ matches among the remaining $(n-1)$ places. The number of ways in which we can have exactly $(r-1)$ matches in the remaining $(n-1)$ places is $(n-1)!\,\beta_{n-1}(r-1)$. Thus

$$P(A_j \cap B_r) = \frac{(n-1)!\,\beta_{n-1}(r-1)}{n!}$$

$$= \frac{p_{n-r}}{(r-1)!\,n}.$$

Hence

$$P(A_j \mid B_r) = \frac{p_{n-r}}{n(r-1)!}\,\frac{r!}{p_{n-r}} = \frac{r}{n}.$$

2.7. Occupancy problems*

A large variety of combinatorial probability problems are equivalent to the problem of distributing n distinct balls into r distinct boxes. Since each of the n balls can go into any one of the r boxes, in all there are r^n different ways of distributing the balls into the boxes. Assuming the balls are distributed at random into the boxes, each of these r^n ways has probability r^{-n}. The underlying probability space Ω therefore has r^n equally likely points. In the problems involving this distribution of balls, we impose various conditions on the occupancy of the boxes and ask for the probability that our stipulated situation occurs. As a first example consider the following problem.

If n balls are distributed at random into r boxes, what is the probability that no box has more than one ball?

To solve the problem note first of all that the required probability is 0 if $n > r$, so assume $n \leq r$. Then (thinking of distributing the balls one by one) the first ball can go in any one of r boxes, the second into any one of the remaining $(r - 1)$ boxes, etc., so in all there are $(r)_n$ different ways. The required probability is then $(r)_n/r^n$.

This probability is exactly the same as that of drawing a sample of size n with replacement from a population of r objects and having all elements in the sample distinct. Also note that r^n is the number of samples of size n from a population of r distinct objects. This is no accident. Random sampling n times with replacement is formally the same as the random distribution of n balls into r boxes. To see this, just think of distributing the balls into the boxes as follows. We first draw a random sample of size n from a set of r objects, and if the ith element in the sample was the jth object we place ball i in box j. It is sometimes useful to think of random sampling with replacement in this manner, i.e., as the random distribution of balls into boxes (see the coupon problem at the end of the chapter).

Consider the random distribution of n balls into r boxes. What is the probability that a specified ball, say ball j, is in a specified box, say box i? If ball j is in box i, then we have $(n - 1)$ more balls to distribute in the r boxes with no restrictions on where they go. Ball j can be placed in box i in only one way, and the $(n - 1)$ remaining balls can be placed into the r boxes in r^{n-1} ways. Thus the required probability is $r^{n-1}/r^n = 1/r$.

Translated into the language of random sampling we see that in a random sample of size n drawn with replacement from a population of r objects, it is equally likely that the jth element in the sample is any one of the r objects.

The above considerations extend easily from one specified box to k boxes, $1 \leq k \leq r$. We leave it as an exercise to show that the probability that k specified balls are in k specified boxes is just r^{-k}. In the language of random sampling this says that if a sample of size n is drawn with replacement from a population of r objects, then the probability that the j_1th, j_2th, ..., j_kth elements in the sample are any k prescribed objects is r^{-k}.

Let $A_j(i)$ be the event that the jth element in the sample is the ith object. Then we have just said that for any choice $j_1 < j_2 < \cdots < j_k$, $1 \le k \le n$, of elements in the sample (i.e., balls) and any choice $i_1, i_2, \ldots, i_k$ of objects (i.e., boxes),

$$P(A_{j_1}(i_1) \cap A_{j_2}(i_2) \cap \cdots \cap A_{j_k}(i_k)) = r^{-k}.$$

Since $P(A_j(i)) = r^{-1}$ for any j and i, we see that

$$P(A_{j_1}(i_1) \cap \cdots \cap A_{j_k}(i_k)) = P(A_{j_1}(i_1)) \cdots P(A_{j_k}(i_k)). \tag{12}$$

Since this is true for all k and all choices of $j_1, \ldots, j_k$, we see that for any $i_1, i_2, \ldots, i_n$ *the events* $A_1(i_1), \ldots, A_n(i_n)$ *are mutually independent.*

If we think of drawing a random sample of size n from a set of r distinct objects as an n-fold repetition of the experiment of choosing one object at random from that set of r distinct objects, then we see that the statement that the events $A_1(i_1), \ldots, A_n(i_n)$ are independent says that the outcome of one experiment has no influence on the outcome of the other experiments. This, of course, is in good accord with our intuitive notion of random sampling.

Example 17. Suppose n balls are distributed at random into r boxes. Find the probability that there are exactly k balls in the first r_1 boxes.

To solve the problem observe that the probability that a given ball is in one of the first r_1 boxes is r_1/r. Think of the distribution of the n balls as an n-fold repetition of the experiment of placing a ball into one of the r boxes. Consider the experiment a success if the ball is placed in one of the first r_1 boxes, and otherwise call it a failure. Then from our results in Section 1.5, we see that the probability that the first r_1 boxes have exactly k balls is

$$\binom{n}{k}\left(\frac{r_1}{r}\right)^k\left(1 - \frac{r_1}{r}\right)^{n-k}.$$

2.8. Number of empty boxes*

We return again to consider the random distribution of n balls into r boxes and seek the probability $p_k(r, n)$ that exactly k boxes are empty.

To begin solving the problem, we let A_i be the event that the ith box is empty. For this event to occur, all of the n balls must be in the remaining $(r - 1)$ boxes, and this can happen in $(r - 1)^n$ ways. Thus $P(A_i) = (r - 1)^n/r^n = (1 - 1/r)^n$.

Similarly, if $1 \le i_1 < i_2 < \cdots < i_k \le r$, then the event $A_{i_1} \cap A_{i_2} \cap \cdots \cap A_{i_k}$ occurs if and only if all of the balls are in the remaining $r - k$ boxes. Consequently, $P(A_{i_1} \cap \cdots \cap A_{i_k}) = (r - k)^n/r^n = (1 - k/r)^n$. We can now apply (7) to compute the probability of $A_1 \cup \cdots \cup A_n$, which is just the event that at least one box is empty. In this situation $S_k = \binom{r}{k}(1 - k/r)^n$, so using (7) we find that

$$P(A_1 \cup \cdots \cup A_r) = \sum_{k=1}^{r} (-1)^{k-1} S_k$$

$$= \sum_{k=1}^{r} (-1)^{k-1} \binom{r}{k} \left(1 - \frac{k}{r}\right)^n.$$

Thus the probability $p_0(r, n)$ that all boxes are occupied is

$$p_0(r, n) = 1 - P(A_1 \cup \cdots \cup A_r) \tag{13}$$

$$= 1 - \sum_{j=1}^{r} (-1)^{j-1} \binom{r}{j} \left(1 - \frac{j}{r}\right)^n$$

$$= \sum_{j=0}^{r} (-1)^j \binom{r}{j} \left(1 - \frac{j}{r}\right)^n.$$

As a next step, let us compute the probability $\alpha_k(r, n)$ that exactly k *specified* boxes (say the first k) are empty. This event can occur only if the n balls are all in the remaining $r - k$ boxes and if none of these $r - k$ boxes are empty. The number of ways we can distribute n balls into $r - k$ boxes in such a manner that no box is empty is $(r - k)^n p_0(r - k, n)$. Thus the required probability is

$$\alpha_k(r, n) = \frac{(r - k)^n p_0(r - k, n)}{r^n} \tag{14}$$

$$= \left(1 - \frac{k}{r}\right)^n p_0(r - k, n).$$

We may now easily compute the probabilities $p_k(r, n)$. For each choice of k distinct numbers $i_1, i_2, \ldots, i_k$ from the set of numbers $\{1, 2, \ldots, n\}$, the event {exactly k boxes $i_1, i_2, \ldots, i_k$ empty} has probability $\alpha_k(r, n)$ and these events are mutually disjoint. There are $\binom{r}{k}$ such events and their union is just the event {exactly k boxes empty}. Thus

$$p_k(r, n) = \binom{r}{k} \left(1 - \frac{k}{r}\right)^n p_0(r - k, n). \tag{15}$$

Using the expression for $p_0(r, n)$ given in (13) we see that

$$p_k(r, n) = \binom{r}{k} \sum_{j=0}^{r-k} (-1)^j \binom{r - k}{j} \left(1 - \frac{j + k}{r}\right)^n. \tag{16}$$

As does the problem of matches, occupancy problems have various reformulations. We mention below one of the more famous of these.

Coupon problem. Coupons or, in the present day, toys are placed in cereal boxes to entice young purchasers. Suppose that there are r different types of coupons or toys, and that a given package is equally likely to contain any one of them. If n boxes are purchased, find the probability of

(a) having collected at least one of each type,
(b) of missing exactly k of the n types.

Exercises

1 The genetic code specifies an amino acid by a sequence of three nucleotides. Each nucleotide can be one of four kinds T, A, C, or G, with repetitions permitted. How many amino acids can be coded in this manner?

2 The Morse code consists of a sequence of dots and dashes with repetitions permitted.
(a) How many letters can be coded for using exactly n symbols?
(b) What is the number of letters that can be coded for using n or fewer symbols?

3 A man has n keys exactly one of which fits the lock. He tries the keys one at a time, at each trial choosing at random from the keys that were not tried earlier. Find the probability that the rth key tried is the correct key.

4 A bus starts with 6 people and stops at 10 different stops. Assuming that passengers are equally likely to depart at any stop, find the probability that no two passengers leave at the same bus stop.

5 Suppose we have r boxes. Balls are placed at random one at a time into the boxes until, for the first time, some box has two balls. Find the probability that this occurs with the nth ball.

6 A box has r balls labeled $1, 2, \ldots, r$. N balls (where $N \leq r$) are selected at random from the box, their numbers noted, and the N balls are then returned to the box. If this procedure is done r times, what is the probability that none of the original N balls are duplicated?

7 If Sam and Peter are among n men who are arranged at random in a line, what is the probability that exactly k men stand between them?

8 A domino is a rectangular block divided into two equal subrectangles as illustrated below. Each subrectangle has a number on it; let these

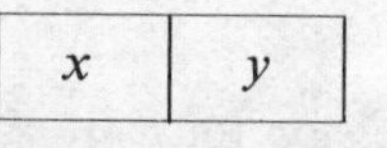

be x and y (not necessarily distinct). Since the block is symmetric, domino (x, y) is the same as (y, x). How many different domino blocks can be made using n different numbers?

9 Consider the problem of matching n objects, and let i and r denote distinct specified positions.

(a) What is the probability that a match occurs at position i and no match occurs at position r?

(b) Given that there is no match at position r what is the probability of a match in position i?

10 Suppose n balls are distributed in n boxes.

(a) What is the probability that exactly one box is empty? *Hint:* use the result of Example 9.

(b) Given that box 1 is empty, what is the probability that only one box is empty?

(c) Given that only one box is empty, what is the probability that box 1 is empty?

11 If n balls are distributed at random into r boxes, what is the probability that box 1 has exactly j balls, $0 \leq j \leq n$?

12 Show that

$$\left(1 - \frac{n-1}{s}\right)^{n-1} \leq \frac{(s)_n}{s^n} \leq \left(1 - \frac{1}{s}\right)^{n-1}.$$

13 A box has b black balls and r red balls. Balls are drawn from the box one at a time without replacement. Find the probability that the first black ball selected is drawn at the nth trial.

The following problem pertains to poker hands. A deck has 52 cards. These cards consist of 4 suits called clubs, diamonds, hearts, and spades. Each suit has 13 cards labeled 2, 3, . . . , 10, J, Q, K, A. A poker hand consists of 5 cards drawn without replacement and without regard to order from the deck. Poker hands of the following types are considered to be in sequence: A, 2, 3, 4, 5; 2, 3, 4, 5, 6; . . . . ; 10, J, Q, K, A.

14 Compute the probability of each of the following poker hands occurring:

(a) Royal flush ((10, J, Q, K, A) of the same suit);

(b) Straight flush (five cards of the same suit in a sequence);

(c) Four of a kind (face values of the form (x, x, x, x, y) where x and y are distinct);

(d) Full house (face values of the form (x, x, x, y, y) where x and y are distinct);

(e) Flush (five cards of the same suit);

(f) Straight (five cards in a sequence, regardless of suit);

(g) Three of a kind (face values of the form (x, x, x, y, z) where x, y, and z are distinct);

(h) Two pairs (face values of the form (x, x, y, y, z) where x, y, and z are distinct);

(i) One pair (face values of the form (w, w, x, y, z) where w, x, y, and z are distinct).

15 A box has 10 balls labeled 1, 2, . . . , 10. Suppose a random sample of size 3 is selected. Find the probability that balls 1 and 6 are among the three selected balls.

16 Cards are dealt from an ordinary deck of playing cards one at a time until the first king appears. Find the probability that this occurs with the nth card dealt.

17 Suppose in a population of r elements a random sample of size n is taken. Find the probability that none of k prescribed elements is in the sample if the method used is
(a) sampling without replacement;
(b) sampling with replacement.

18 Suppose a random sample of size n is drawn from a population of r objects without replacement. Find the probability that k given objects are included in the sample.

19 Suppose n objects are permuted at random among themselves. Prove that the probability that k specified objects occupy k specified positions is $(n - k)!/n!$.

20 With reference to Example 14, show that

$$\frac{\binom{n}{k}\binom{r-n}{m-k}}{\binom{r}{m}} = \frac{\binom{m}{k}\binom{r-m}{n-k}}{\binom{r}{n}}.$$

21 A box contains 40 good and 10 defective fuses. If 10 fuses are selected, what is the probability they will all be good?

22 What is the probability that the bridge hands of north and south together (a total of 26 cards) contain exactly 3 aces?

23 What is the probability that if 4 cards are drawn from a deck, 2 will be black and 2 will be red?

24 Find the probability that a poker hand of 5 cards will contain no card smaller than 7, given that it contains at least 1 card over 10, where aces are treated as high cards.

25 If you hold 3 tickets to a lottery for which n tickets were sold and 5 prizes are to be given, what is the probability that you will win at least 1 prize?

26 A box of 100 washers contains 5 defective ones. What is the probability that two washers selected at random (without replacement) from the box are both good?

27 Two boxes each have r balls labeled 1, 2, . . . , r. A random sample of size $n \leq r$ is drawn without replacement from each box. Find the probability that the samples contain exactly k balls having the same numbers in common.

3 Discrete Random Variables

Consider the experiment of tossing a coin three times where the probability of a head on an individual toss is p. Suppose that for each toss that comes up heads we win \$1, but for each toss that comes up tails we lose \$1. Clearly, a quantity of interest in this situation is our total winnings. Let X denote this quantity. It is clear that X can only be one of the values \$3, \$1, −\$1, and −\$3. We cannot with certainty say which of these values X will be, since that value depends on the outcome of our random experiment. If for example the outcome is HHH, then X will be \$3; while for the outcome HTH, X will be \$1. In the following table we list the values of X (in dollars) corresponding to each of the eight possible outcomes.

ω	$X(\omega)$	$P\{\omega\}$
HHH	3	p^3
HHT	1	$p^2(1-p)$
HTH	1	$p^2(1-p)$
THH	1	$p^2(1-p)$
HTT	-1	$p(1-p)^2$
THT	-1	$p(1-p)^2$
TTH	-1	$p(1-p)^2$
TTT	-3	$(1-p)^3$

We can think of X as a real-valued function on the probability space corresponding to the experiment. For each $\omega \in \Omega$, $X(\omega)$ is then one of the values, 3, 1, −1, −3. Consider, for example, the event $\{\omega: X(\omega) = 1\}$. This set contains the three points ω_2, ω_3, and ω_4 corresponding to the outcomes HHT, HTH, and THH, respectively. The last column in the table gives the probabilities associated with the eight possible outcomes of our experiment. From that table we see that the event $\{\omega: X(\omega) = 1\}$ has probability $3p^2(1-p)$. We usually abbreviate this by saying that $\{X = 1\}$ has probability $3p^2(1 - p)$. Similar considerations, of course, apply to the other values that X assumes. We see, therefore, that for each possible value of X, there is a precisely defined probability for X assuming that value. As we shall see in the next section, the quantity X is an example of what is called a discrete random variable.

3.1. Definitions

Let $(\Omega, \mathscr{A}, P)$ be an arbitrary probability space, and let X be a real-valued function on Ω taking only a finite or countably infinite number of values $x_1, x_2, \ldots$. As in the example just given, we would certainly like to be able to talk about the probability that X assumes the value x_i, for each i. For this to be the case we need to know that for each i, $\{\omega \in \Omega : X(\omega) = x_i\}$ is an event, i.e., is a member of $\mathscr{A}$. If, as in the previous example, $\mathscr{A}$ is the σ-field of all subsets of Ω then this is certainly the case. For in that case, no matter what x_i might be, $\{\omega : X(\omega) = x_i\}$ is a subset of Ω and hence a member of $\mathscr{A}$, since $\mathscr{A}$ contains every possible subset of Ω. However, as was indicated in Section 1.2, in general $\mathscr{A}$ does not consist of all subsets of Ω, so we have no guarantee that $\{\omega \in \Omega : X(\omega) = x_i\}$ is in $\mathscr{A}$. The only reasonable way out is to *explicitly assume* that X is a function on Ω such that this desired property holds. This leads us to the following.

Definition 1 *A discrete real-valued random variable X on a probability space $(\Omega, \mathscr{A}, P)$ is a function X with domain Ω and range a finite or countably infinite subset $\{x_1, x_2, \ldots\}$ of the real numbers R such that $\{\omega : X(\omega) = x_i\}$ is an event for all i.*

By definition, then, $\{\omega : X(\omega) = x_i\}$ is an event so we can talk about its probability. For brevity we usually write the event $\{\omega : X(\omega) = x_i\}$ as $\{X = x_i\}$ and denote the probability of this event as $P(X = x_i)$ rather than $P(\{\omega : X(\omega) = x_i\})$.

Let X be a discrete real-valued random variable. Then for any real number x, $\{\omega : X(\omega) = x\}$ is an event. Indeed, if $x_1, x_2, \ldots$ are the values that X can assume, then $\{\omega : X(\omega) = x_i\}$ is an event by the definition of a discrete real-valued random variable. If x is not one of these numbers, then $\{\omega : X(\omega) = x\} = \varnothing$, which is also an event.

If the possible values of a discrete random variable X consist only of integers or of nonnegative integers, we say that X is respectively an integer-valued random variable or a nonnegative integer-valued random variable. Most of the discrete random variables that arise in applications are nonnegative integer-valued.

Definition 2 *The real-valued function f defined on R by $f(x) = P(X = x)$ is called the discrete density function of X. A number x is called a possible value of X if $f(x) > 0$.*

Whenever necessary, we will denote the density function of X by f_X to show that it is the density function for the random variable X.

Example 1. Let X be the random variable introduced at the beginning of this chapter in our discussion of tossing a coin three times with, say, $p = .4$. Then X has the discrete density f given by

$$f(-3) = .216, \qquad f(-1) = .432, \qquad f(1) = .288, \qquad f(3) = .064,$$

and $f(x) = 0$ if $x \neq -3, -1, 1, 3$. This density can be represented in terms of a diagram as illustrated in Figure 1.

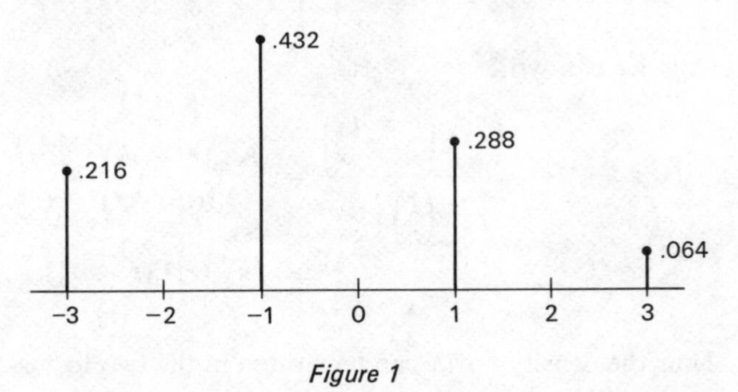

Figure 1

Example 2. Binomial distribution. Consider n independent repetitions of the simple success-failure experiment discussed in Section 1.5. Let S_n denote the number of successes in the n trials. Then S_n is a random variable that can only assume the values $0, 1, 2, \ldots, n$. In Chapter 1 we showed that for the integer k, $0 \leq k \leq n$,

$$P(S_n = k) = \binom{n}{k} p^k(1 - p)^{n-k};$$

hence the density f of S_n is given by

$$f(x) = \begin{cases} \binom{n}{x} p^x(1 - p)^{n-x}, & x = 0, 1, 2, \ldots, n, \\ 0, & \text{elsewhere.} \end{cases}$$

This density, which is among the most important densities that occur in probability theory, is called the *binomial density* with parameters n and p. The density from Example 1 is a binomial density with parameters $n = 3$ and $p = .4$.

One often refers to a random variable X having a binomial density by saying that X has a binomial *distribution* (with parameters n and p if one wants to be more precise). Similar phraseology is also used for other random variables having a named density.

As explained in Chapter 2, the binomial distribution arises in random sampling with replacement. For random sampling without replacement we have the following.

Example 3. Hypergeometric distribution. Consider a population of r objects, of which r_1 are of one type and $r_2 = r - r_1$ are of a second type. Suppose a random sample of size $n \le r$ is drawn from the population. Let X be the number of objects of the first type in the sample. Then X is a random variable whose possible values are $0, 1, 2, \ldots, n$. From the results in Section 2.4 we know that

$$P(X = x) = \frac{\binom{r_1}{x}\binom{r - r_1}{n - x}}{\binom{r}{n}}, \qquad x = 0, 1, 2, \ldots, n.$$

Now we can write

$$\frac{\binom{r_1}{x}\binom{r - r_1}{n - x}}{\binom{r}{n}} = \frac{(r_1)_x(r - r_1)_{n-x}}{x!(n - x)!} \frac{n!}{(r)_n}$$

$$= \binom{n}{x} \frac{(r_1)_x(r - r_1)_{n-x}}{(r)_n}.$$

Thus the density f of X can be written in the two forms

$$f(x) = \begin{cases} \dfrac{\binom{r_1}{x}\binom{r - r_1}{n - x}}{\binom{r}{n}}, & x = 0, 1, 2, \ldots, n, \\ 0, & \text{elsewhere} \end{cases}$$

or

$$f(x) = \begin{cases} \binom{n}{x} \dfrac{(r_1)_x(r - r_1)_{n-x}}{(r)_n}, & x = 0, 1, 2, \ldots, n, \\ 0, & \text{elsewhere.} \end{cases}$$

This density is called the *hypergeometric density*.

Here are a few more examples of random variables.

Example 4. Constant random variable. Let c be a real number. Then the function X defined by $X(\omega) = c$ for all ω is a discrete random variable, since the set $\{\omega: X(\omega) = c\}$ is the entire set Ω and Ω is an event. Clearly, $P(X = c) = 1$, so the density f of X is simply $f(c) = 1$ and $f(x) = 0$, $x \ne c$. Such a random variable is called a constant random variable. It is from this point of view that a numerical constant is considered a random variable.

Example 5. Indicator random variable. Let A be an event. Set $X(\omega) = 1$ if $\omega \in A$ and $X(\omega) = 0$ if $\omega \notin A$. Then the event A occurs if and only if $X = 1$. This random variable is called the indicator random variable of A because the value of X tells whether or not the event A occurs.

Conversely, if X is a random variable on a probability space $(\Omega, \mathscr{A}, P)$ taking the values 1 or 0, then X is the indicator random variable of the event

$$A = \{\omega: X(\omega) = 1\}.$$

Let $p = P(X = 1)$. The density f of X is then given by

$$f(0) = 1 - p, \qquad f(1) = p, \qquad \text{and} \qquad f(x) = 0, \qquad x \neq 0 \text{ or } 1.$$

Example 6. Consider the following game of chance. A circular target of radius 1 is zoned into n concentric disks of radius $1/n, 2/n, \ldots, n/n = 1$, as illustrated in Figure 2 for the case $n = 5$. A dart is tossed at random onto the circle, and if it lands in the annular zone between the circles with radii i/n and $(i + 1)/n$, $n - i$ dollars are won, $i = 0, 1, 2, \ldots, n - 1$. Let X denote the amount of money won. Find the density of X.

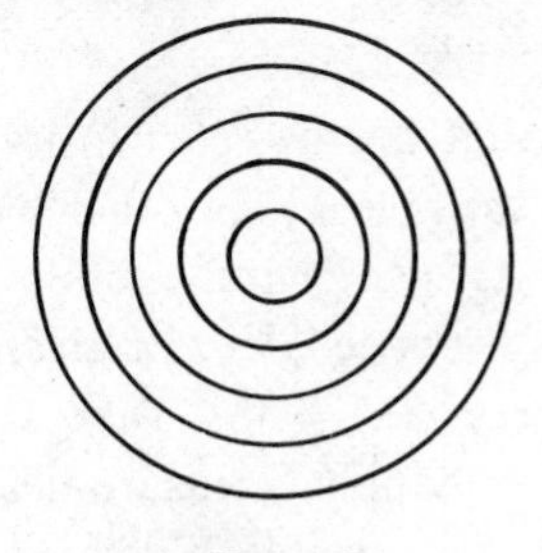

Figure 2

The probability space for this experiment will be chosen to be the uniform probability space on the disk of radius 1. Clearly X is a discrete random variable on this space with the possible values $1, 2, \ldots, n$. The event $A = \{X = n - i\}$ occurs if and only if the dart lands in the region bounded by the circles of radii i/n and $(i + 1)/n$. According to our discussion in Section 1.2 the probability of A is the area of A divided by the area of the unit disk. Thus for $i = 0, 1, 2, \ldots, n - 1$

$$\begin{aligned} P(X = n - i) &= P(A) \\ &= \frac{\pi\left[\left(\frac{i+1}{n}\right)^2 - \left(\frac{i}{n}\right)^2\right]}{\pi} = \frac{2i + 1}{n^2}. \end{aligned}$$

Setting $n - i = x$ we see that the density of X is

$$f(x) = \begin{cases} \dfrac{2(n - x) + 1}{n^2}, & x = 1, 2, \ldots, n, \\ 0, & \text{elsewhere.} \end{cases}$$

The density f of a discrete random variable X has the following three important properties:

(i) $f(x) \geq 0, x \in R$.

(ii) $\{x: f(x) \neq 0\}$ *is a finite or countably infinite subset of* R. *Let* $\{x_1, x_2, \ldots\}$ *denote this set. Then*

(iii) $\sum_i f(x_i) = 1$.

Properties (i) and (ii) are immediate from the definition of the discrete density function of X. To see that (iii) holds, observe that the events $\{\omega: X(\omega) = x_i\}$ are mutually disjoint and their union is Ω. Thus

$$\sum_i f(x_i) = \sum_i P(X = x_i)$$

$$= P\left(\bigcup_i \{X = x_i\}\right) = P(\Omega) = 1.$$

Definition 3 *A real-valued function f defined on R is called a discrete density function provided that it satisfies properties* (i), (ii), *and* (iii) *stated above.*

It is easy to see that any discrete density function f is the density function of some random variable X. In other words, given f we can construct a probability space $(\Omega, \mathscr{A}, P)$ and a random variable X defined on Ω whose discrete density is f. Indeed, let f be given and suppose $\{x_1, x_2, \ldots\}$ is the set of values where $f(x) \neq 0$. Take $\Omega = \{x_1, x_2, \ldots\}$, $\mathscr{A}$ all subsets of Ω, and P the probability measure defined on $\mathscr{A}$ by $P(\{\omega\}) = f(x_i)$ if $\omega = x_i$. The random variable X defined by $X(\omega) = x_i$ if $\omega = x_i$ is then such a random variable. To see this note that $\{\omega: X(\omega) = x_i\} = \{x_i\}$ and thus

$$P(X = x_i) = P(\{x_i\}) = f(x_i).$$

The above result assures us that statements like "Let X be a random variable with discrete density f" always make sense, even if we do not specify directly a probability space upon which X is defined. To save writing we will henceforth use the term density instead of discrete density throughout the remainder of this chapter.

The notion of a discrete random variable forms a convenient way of describing a random experiment that has a finite or countably infinite number of possible outcomes. We need not bother to set up a probability space for the experiment. Instead we can simply introduce a random variable X taking values $x_1, x_2, \ldots$ such that $X = x_i$ if and only if the experiment results in the ith outcome. Thus, for example, in drawing a card at random from a deck of n cards, we can let $X = i$ if the ith card was drawn. Then $P(X = i) = n^{-1}$, so we could describe the experiment by saying we observe a random variable X taking integer values $1, 2, \ldots, n$ and having $f(x) = n^{-1}$ for $x = 1, 2, \ldots, n$, and $f(x) = 0$ elsewhere for its density function.

In general, conducting an experiment which has a finite or countably infinite number of possible outcomes can be described as observing the value of a discrete random variable X. Many times, in fact, this is how the experiment already appears to us, and often it is easier to think of the experiment in these terms rather than in terms of a probability space.

As an illustration of this idea consider the experiment of picking a point at random from the finite subset S of R consisting of the distinct points $x_1, x_2, \ldots, x_s$. Then the function f defined by

$$f(x) = \begin{cases} s^{-1}, & x = x_1, x_2, \ldots, x_s, \\ 0, & \text{elsewhere} \end{cases}$$

is clearly a discrete density function. A random variable X having this density is said to be *uniformly distributed on* S. Observing a value of X corresponds to our intuitive notion of choosing a point at random from S.

We will now introduce two more discrete densities that are very useful for solving certain classes of problems whose importance will become apparent later.

Example 7. Geometric densities. Let $0 < p < 1$. Then the real-valued function f defined on R by

$$f(x) = \begin{cases} p(1 - p)^x, & x = 0, 1, 2, \ldots, \\ 0, & \text{elsewhere} \end{cases}$$

is a discrete density function called the *geometric density* with parameter p.

To see that f is a density, all that needs to be checked is that condition (iii) holds, for here conditions (i) and (ii) are obviously satisfied. But (iii) follows from the familiar fact that the sum of the geometric series $\sum_{x=0}^{\infty} (1 - p)^x$ is just p^{-1}.

Example 8. Negative binomial densities. Let α be any positive real number and let $0 < p < 1$. A density closely related to the geometric is the *negative binomial density* with parameters α and p defined by

$$f(x) = \begin{cases} p^\alpha \binom{-\alpha}{x} (-1)^x (1 - p)^x, & x = 0, 1, 2, \ldots, \\ 0, & \text{elsewhere}. \end{cases} \tag{1}$$

To show that this is a density we must verify that properties (i)–(iii) hold. Here property (ii) is obviously true. That (i) holds may be seen as follows. For x a nonnegative integer,

$$\begin{aligned} \binom{-\alpha}{x} &= \frac{(-\alpha)_x}{x!} \\ &= \frac{(-\alpha)(-\alpha - 1)\cdots(-\alpha - x + 1)}{x!} \end{aligned}$$

$$= \frac{(-1)^x(\alpha)(\alpha + 1)\cdots(\alpha + x - 1)}{x!}$$

$$= (-1)^x \frac{(\alpha + x - 1)_x}{x!}$$

$$= (-1)^x \binom{\alpha + x - 1}{x}.$$

Thus

$$(2) \qquad p^\alpha \binom{-\alpha}{x} (-1)^x(1 - p)^x = p^\alpha \binom{\alpha + x - 1}{x} (1 - p)^x.$$

Since the right-hand side of (2) is clearly nonnegative we see that (i) holds. To verify (iii), recall that the Taylor series of $(1 - t)^{-\alpha}$ for $-1 < t < 1$ is

$$(3) \qquad (1 - t)^{-\alpha} = \sum_{x=0}^{\infty} \binom{-\alpha}{x} (-t)^x.$$

From (3) with $t = 1 - p$, we see that

$$p^{-\alpha} = \sum_{x=0}^{\infty} \binom{-\alpha}{x} (-1)^x(1 - p)^x$$

and hence that $\sum_x f(x) = 1$.

From (2) we see that we can write the negative binomial density in the alternate form

$$(4) \qquad f(x) = \begin{cases} p^\alpha \dbinom{\alpha + x - 1}{x} (1 - p)^x, & x = 0, 1, 2, \ldots, \\ 0, & \text{elsewhere.} \end{cases}$$

For some purposes this form is more useful than that given in (1). Observe that the geometric density with parameter p is a negative binomial density with parameters $\alpha = 1$ and p.

Example 9. Poisson densities. Let λ be a positive number. The Poisson density with parameter λ is defined as

$$f(x) = \begin{cases} \dfrac{\lambda^x e^{-\lambda}}{x!}, & x = 0, 1, 2, \ldots, \\ 0, & \text{elsewhere.} \end{cases}$$

It is obvious that this function satisfies properties (i) and (ii) in the definition of a discrete density function. Property (iii) follows immediately from the Taylor series expansion of the exponential function, namely,

$$e^\lambda = \sum_{x=0}^{\infty} \frac{\lambda^x}{x!}.$$

Many counting type random phenomena are known from experience to be approximately Poisson distributed. Some examples of such phenomena are the number of atoms of a radioactive substance that disintegrate in a unit time interval, the number of calls that come into a telephone exchange in a unit time interval, the number of misprints on a page of a book, and the number of bacterial colonies that grow on a petri dish that has been smeared with a bacterial suspension. A full treatment of these models requires the notion of a Poisson process, which will be discussed in Chapter 9.

3.2. Computations with densities

So far we have restricted our attention to computing $P(X = x)$. Often we are interested in computing the probability of $\{\omega: X(\omega) \in A\}$ where A is some subset of R other than a one-point set.

Let A be any subset of R and let X be a discrete random variable having distinct possible values $x_1, x_2, \ldots$. Then $\{\omega: X(\omega) \in A\}$ is an event. To see this, observe that

$$\{\omega \mid X(\omega) \in A\} = \bigcup_{x_i \in A} \{\omega \mid X(\omega) = x_i\}, \tag{5}$$

where by $\bigcup_{x_i \in A}$ we mean the union over all i such that $x_i \in A$. Usually the event $\{\omega: X(\omega) \in A\}$ is abbreviated to $\{X \in A\}$, and its probability is denoted by $P(X \in A)$. If $-\infty \leq a < b \leq \infty$ and A is an interval with end points a and b, say $A = (a, b]$, then we usually write $P(a < X \leq b)$ instead of $P(X \in (a, b])$. Similar notation is used for the other intervals with these endpoints.

An abbreviated notation is also used for conditional probabilities. Thus, for example, if A and B are two subsets of R we write $P(X \in A \mid X \in B)$ for the conditional probability of the event $\{X \in A\}$ given the event $\{X \in B\}$.

Let f be the density of X. We can compute $P(X \in A)$ directly from the density f by means of the formula

$$P(X \in A) = \sum_{x_i \in A} f(x_i), \tag{6}$$

where by $\sum_{x_i \in A}$ we mean the sum over all i such that $x_i \in A$. This formula follows immediately from (5) since the events $\{\omega \mid X(\omega) = x_i\}$, $i = 1, 2, \ldots$, are disjoint. The right side of (6) is usually abbreviated as $\sum_{x \in A} f(x)$. In terms of this notation (6) becomes

$$P(X \in A) = \sum_{x \in A} f(x). \tag{7}$$

The function $F(t)$, $-\infty < t < \infty$, defined by

$$F(t) = P(X \leq t) = \sum_{x \leq t} f(x), \qquad -\infty < t < \infty,$$

is called the *distribution function* of the random variable X or of the density f. It follows immediately from the definition of the distribution function that

$$P(a < X \leq b) = P(X \leq b) - P(X \leq a) = F(b) - F(a).$$

If X is an integer-valued random variable, then

$$F(t) = \sum_{x=-\infty}^{[t]} f(x),$$

where $[t]$ denotes the greatest integer less than or equal to t (e.g., $[2.6] = [2] = 2$). We see that F is a nondecreasing function and that, for any integer x, F has a jump of magnitude $f(x)$ at x and F is constant on the interval $[x, x + 1)$. Further properties of distribution functions will be obtained, from a more general viewpoint, in Chapter 5.

Example 10. Set $S = \{1, 2, \ldots, 10\}$ and let X be uniformly distributed on S. Then $f(x) = 1/10$ for $x = 1, 2, \ldots, 10$ and $f(x) = 0$, elsewhere. The distribution function of X is given by $F(t) = 0$ for $t < 1$, $F(t) = 1$ for $t > 10$ and

$$F(t) = \sum_{x=1}^{[t]} f(x) = \frac{[t]}{10}, \qquad 1 \leq x \leq 10.$$

A graph of this distribution function is given in Figure 3. The probability $P(3 < X \leq 5)$ can be calculated either as

$$P(3 < X \leq 5) = f(4) + f(5) = 2/10$$

or as

$$P(3 < X \leq 5) = F(5) - F(3) = 5/10 - 3/10 = 2/10.$$

Similarly $P(3 \leq X \leq 5)$ is obtained as

$$P(3 \leq X \leq 5) = f(3) + f(4) + f(5) = 3/10$$

or as

$$P(3 \leq X \leq 5) = P(2 < X \leq 5) = F(5) - F(2) = 5/10 - 2/10 = 3/10.$$

Figure 3

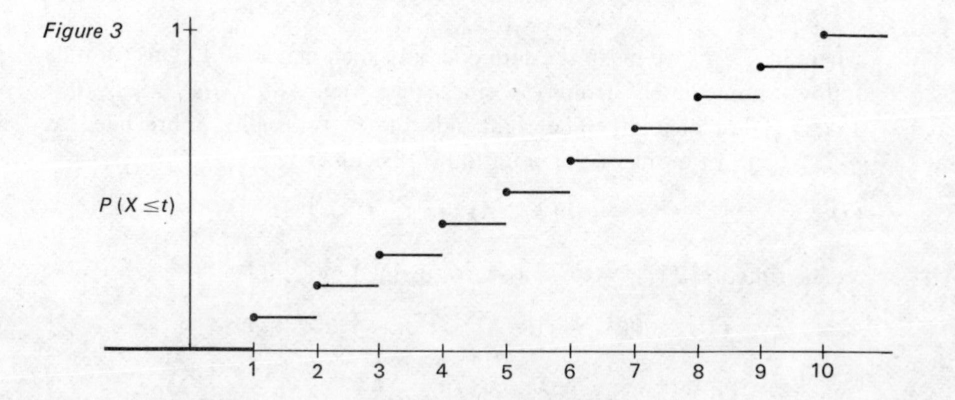

Example 11. Let X have a geometric distribution with parameter p. Find the distribution function of X and find $P(X \geq x)$ for the nonnegative integer x.

The density of X, according to Example 7, is

$$f(x) = \begin{cases} p(1-p)^x, & x = 0, 1, 2, \ldots, \\ 0, & \text{elsewhere.} \end{cases}$$

Thus $F(t) = 0$ for $t < 0$ and

$$F(t) = \sum_{x=0}^{[t]} p(1-p)^x, \qquad t \geq 0.$$

Using the formula for the sum of a finite geometric progression we conclude that

$$F(t) = 1 - (1-p)^{[t]+1}, \qquad t \geq 0.$$

In particular, for x a nonnegative integer, $F(x-1) = 1 - (1-p)^x$ and hence

$$\begin{aligned} P(X \geq x) &= 1 - P(X < x) = 1 - P(X \leq x - 1) \\ &= 1 - F(x-1) = (1-p)^x. \end{aligned}$$

Geometrically distributed random variables arise naturally in applications. Suppose we have a piece of equipment, such as an electrical fuse, that neither deteriorates nor improves in the course of time but can fail due to sporadic chance happenings that occur homogeneously in time. Let the object be observed at fixed time periods such as hours or days, and let X be the number of time units up to and including the first failure, assuming that the object is new at time 0. Clearly X is a discrete random variable whose possible values are found among the integers $1, 2, 3, \ldots$. The event $\{X = n\}$ occurs if and only if the object first fails at the nth time period. Our intuitive notion that the object neither deteriorates nor improves with time can be precisely formulated as follows. If we know that the object has not failed by time n, i.e., the first failure is after time n so $X > n$, then the probability that it does not fail until after time $n + m$, i.e., $P(X > n + m \mid X > n)$, should be the same as the probability of starting with an object which is new at time n and having it not fail until after time $n + m$. The fact that the failure causes occur homogeneously in time can be taken to mean that this probability depends only on the number of time periods that elapse between n and $n + m$, namely m, but not on n. Thus $P(X > n)$ should satisfy the equation

$$P(X > n + m \mid X > n) = P(X > m). \tag{8}$$

Since

$$P(X > n + m \mid X > n) = \frac{P(X > n + m)}{P(X > n)},$$

we can rewrite (8) as

(9) $P(X > n + m) = P(X > n)P(X > m), \qquad n, m = 0, 1, 2, \ldots.$

Setting $n = m = 0$ we see that $P(X > 0) = P(X > 0)^2$, so $P(X > 0)$ equals 1 or 0. If $P(X > 0) = 0$, then $P(X = 0) = 1$, which is impossible in our case since X can assume only values that are positive integers. Therefore $P(X > 0) = 1$.

Set $p = P(X = 1)$. Then $P(X > 1) = 1 - p$ and from (9) we see that

$$P(X > n + 1) = (1 - p)P(X > n).$$

By iteration on n it follows that $P(X > n) = (1 - p)^n$. Thus for $n = 1, 2, \ldots,$

(10) $$P(X = n) = P(X > n - 1) - P(X > n)$$
$$= (1 - p)^{n-1} - (1 - p)^n = p(1 - p)^{n-1}.$$

If $p = 0$ then $P(X = n) = 0$ for all $n = 1, 2, \ldots$ and thus $P(X = +\infty) = 1$, i.e., the object never fails. We exclude this case from consideration. Likewise $p = 1$ is excluded because then $P(X = 1) = 1$, so the object always fails.

Let $Y = X - 1$. Then Y assumes the values $0, 1, 2, \ldots$ with probabilities $P(Y = n) = p(1 - p)^n$. We see therefore that Y has the geometric distribution with parameter p.

As we have just shown, the random variable $Y = X - 1$ is geometrically distributed. This example is typical in the sense that geometrically distributed random variables usually arise in connection with the waiting time for some event to occur. We shall discuss this in more detail after we treat independent trials in Section 3.4.

3.3. Discrete random vectors

It often happens that we are interested in studying the relationship between two or more random variables. Thus, for example, in drawing a random sample of size n from a box of r balls labeled $1, 2, \ldots, r$, we might want to know the largest number Y on the balls selected as well as the smallest number Z on the selected balls.

Let $(\Omega, \mathscr{A}, P)$ be a probability space and let $X_1, X_2, \ldots, X_r$ be r discrete random variables defined on this space. Then for each point $\omega \in \Omega$ each of the random variables $X_1, \ldots, X_r$ takes on one of its possible values, which will be indicated by writing

$$X_1(\omega) = x_1, X_2(\omega) = x_2, \ldots, X_r(\omega) = x_r.$$

Instead of thinking of observing r real numbers $x_1, x_2, \ldots, x_r$ we can think of observing *one* r-tuple $\mathbf{x} = (x_1, x_2, \ldots, x_r)$, where for each index

i, x_i is one of the finite or countably infinite number of values that the random variable X_i can assume.

Let R^r denote the collection of all r-tuples of real numbers. A point $\mathbf{x} = (x_1, \ldots, x_r)$ of R^r is usually called an *r-dimensional vector*. Thus for each $\omega \in \Omega$, the r values $X_1(\omega), \ldots, X_r(\omega)$ define a point

$$\mathbf{X}(\omega) = (X_1(\omega), X_2(\omega), \ldots, X_r(\omega))$$

of R^r. This defines an r-dimensional vector-valued function on Ω, $\mathbf{X}: \Omega \to R^r$, which is usually written $\mathbf{X} = (X_1, X_2, \ldots, X_r)$. The function $\mathbf{X}$ is called a *discrete r-dimensional random vector*.

We have just defined an r-dimensional random vector in terms of r real-valued random variables. Alternatively, an r-dimensional random vector can be defined directly as a function $\mathbf{X}: \Omega \to R^r$ by extending the definition of a real-valued random variable almost verbatim.

Definition 4 *A discrete r-dimensional random vector* $\mathbf{X}$ *is a function* $\mathbf{X}$ *from* Ω *to* R^r *taking on a finite or countably infinite number of values* $\mathbf{x}_1, \mathbf{x}_2, \ldots$ *such that*

$$\{\omega: \mathbf{X}(\omega) = \mathbf{x}_i\}$$

is an event for all i.

The discrete density function f for the random vector $\mathbf{X}$ is defined by

$$f(x_1, \ldots, x_r) = P(X_1 = x_1, \ldots, X_r = x_r)$$

or equivalently

$$f(\mathbf{x}) = P(\mathbf{X} = \mathbf{x}), \qquad \mathbf{x} \in R^r.$$

The probability that $\mathbf{X}$ belongs to the subset A of R^r can be found by using the analog of (7), namely,

$$P(\mathbf{X} \in A) = \sum_{\mathbf{x} \in A} f(\mathbf{x}).$$

As in the one-dimensional case, this function f has the following three properties:

(i) $f(\mathbf{x}) \geq 0$, $\mathbf{x} \in R^r$.

(ii) $\{\mathbf{x}: f(\mathbf{x}) \neq 0\}$ *is a finite or countably infinite subset of* R^r, *which will be denoted by* $\{\mathbf{x}_1, \mathbf{x}_2, \ldots\}$.

(iii) $\sum_i f(\mathbf{x}_i) = 1$.

Any real-valued function f defined on R^r having these three properties will be called a discrete r-dimensional density function. The argument given in the one-dimensional case applies verbatim to show that any r-dimensional discrete density function is the density function of some r-dimensional random vector.

There is a certain amount of traditional terminology that goes along with random vectors and their density functions. Let $\mathbf{X} = (X_1, X_2, \ldots, X_r)$ be an r-dimensional random vector with density f. Then the function f is usually called the *joint density* of the random variables $X_1, X_2, \ldots, X_r$. The density function of the random variable X_i is then called the ith *marginal density* of $\mathbf{X}$ or of f.

Let X and Y be two discrete random variables. For any real numbers x and y the set $\{\omega \mid X(\omega) = x \text{ and } Y(\omega) = y\}$ is an event that we will usually denote by $\{X = x, Y = y\}$. Suppose that the distinct possible values of X are $x_1, x_2, \ldots$, and that the distinct possible values of Y are $y_1, y_2, \ldots$. For each x, the events $\{X = x, Y = y_j\}$, $j = 1, 2, \ldots$, are disjoint and their union is the event $\{X = x\}$. Thus

$$\begin{aligned} P(X = x) &= P\left(\bigcup_j \{X = x, Y = y_j\}\right) \\ &= \sum_j P(X = x, Y = y_j) = \sum_y P(X = x, Y = y). \end{aligned}$$

This last expression results from using the same notational convention that was introduced for random variables in Section 3.2. Similarly,

$$\begin{aligned} P(Y = y) &= P\left(\bigcup_i \{X = x_i, Y = y\}\right) \\ &= \sum_i P(X = x_i, Y = y) = \sum_x P(X = x, Y = y). \end{aligned}$$

In other words, if we know the joint density of two discrete random variables X and Y then we can compute the density f_X of X by summing over y and the density f_Y of Y by summing over x. Thus, in terms of densities, if f is the joint density of X and Y, then

$$f_X(x) = \sum_y f(x, y) \tag{11}$$

and

$$f_Y(y) = \sum_x f(x, y). \tag{12}$$

Example 12. Suppose two cards are drawn at random without replacement from a deck of 3 cards numbered 1, 2, 3. Let X be the number on the first card and let Y be the number on the second card. Then the joint density f of X and Y is given by $f(1, 2) = f(1, 3) = f(2, 1) = f(2, 3) = f(3, 1) = f(3, 2) = 1/6$ and $f(x, y) = 0$ elsewhere. The first marginal density, that is, the density of X is given by

$$f_X(1) = f(1, 1) + f(1, 2) + f(1, 3) = 0 + 1/6 + 1/6 = 2/6 = 1/3$$

and similarly for $x = 2$ and 3. Therefore $f_X(x) = 1/3$, $x = 1, 2, 3$, and $f_X(x) = 0$ elsewhere, as it should be.

Example 13. Suppose X and Y are random variables that assume the values x and y, where $x = 1$ or 2 and $y = 1, 2, 3, 4$, with probabilities given by the following table.

$x \backslash y$	1	2	3	4
1	1/4	1/8	1/16	1/16
2	1/16	1/16	1/4	1/8

Then $f_X(1) = \sum_{y=1}^{4} f(1, y) = 1/4 + 1/8 + 1/16 + 1/16 = 1/2$, and $f_X(2) = 1 - f_X(1) = 1/2$ so X has the uniform distribution on 1, 2. Similarly

$$f_Y(1) = 1/4 + 1/16 = 5/16,\quad f_Y(2) = 3/16,\quad f_Y(3) = 5/16,\quad f_Y(4) = 3/16.$$

3.4. Independent random variables

Consider the experiment of tossing a coin and rolling a die. Intuitively, we believe that whatever the outcome of the coin toss is, it should have no influence on the outcome of the die roll, and vice-versa. We now wish to construct a probability model that reflects these views. Let X be a random variable that is 1 or 0 according as the coin lands heads or tails, i.e., such that the event $\{X = 1\}$ represents the outcome that the coin lands heads and the event $\{X = 0\}$ represents the outcome that the coin lands tails. In a similar way we represent the outcome of the die roll by a random variable Y that takes the value $1, 2, \ldots,$ or 6 according as the die roll results in the face number $1, 2, \ldots,$ or 6. The outcome of the combined experiment can then be given by the random vector (X, Y). Our intuitive notion that the outcome of the coin toss and die roll have no influence on each other can be stated precisely by saying that if x is one of the numbers 1 or 0 and y is one of the numbers $1, 2, \ldots, 6$, then the events $\{X = x\}$ and $\{Y = y\}$ should be independent. Thus, the random vector (X, Y) should have the joint density $f(x, y)$ given by

$$f(x, y) = \begin{cases} P(X = x)P(Y = y), & x = 0, 1, \quad y = 1, 2, \ldots, 6, \\ 0, & \text{elsewhere.} \end{cases}$$

In other words the joint density f of X and Y should be given by

$$f(x, y) = f_X(x)f_Y(y).$$

Definition 5 *Let $X_1, X_2, \ldots, X_r$ be r discrete random variables having densities $f_1, f_2, \ldots, f_r$ respectively. These random variables* *are said to be mutually independent if their joint density function f is given by*

$$f(x_1, x_2, \ldots, x_r) = f_1(x_1)f_2(x_2)\cdots f_r(x_r). \tag{13}$$

The random variables are said to be dependent if they are not independent. As in the case of the combined experiment of tossing a coin and rolling a die, the notion of independent random variables forms a convenient way to precisely formulate our intuitive notions that experiments are independent of each other.

Consider two independent discrete random variables having densities f_X and f_Y, respectively. Then for any two subsets A and B of R

$$P(X \in A, Y \in B) = P(X \in A)P(Y \in B). \tag{14}$$

To see this, note that

$$\begin{aligned} P(X \in A, Y \in B) &= \sum_{x \in A} \sum_{y \in B} f_{X,Y}(x, y) \\ &= \sum_{x \in A} \sum_{y \in B} f_X(x) f_Y(y) \\ &= \left[\sum_{x \in A} f_X(x)\right]\left[\sum_{y \in B} f_Y(y)\right] \\ &= P(X \in A)P(Y \in B). \end{aligned}$$

Formula (14) above extends easily from 2 to r independent random variables. Thus if $A_1, A_2, \ldots, A_r$ are any r subsets of R then

$$P(X_1 \in A_1, \ldots, X_r \in A_r) = P(X_1 \in A_1) \cdots P(X_r \in A_r). \tag{15}$$

Example 14. Let X and Y be independent random variables each geometrically distributed with parameter p.

(a) Find the distribution of $\min(X, Y)$.

(b) Find $P(\min(X, Y) = X) = P(Y \geq X)$.

(c) Find the distribution of $X + Y$.

(d) Find $P(Y = y \mid X + Y = z)$ for $y = 0, 1, \ldots, z$.

To solve (a) we observe that for z a nonnegative integer

$$P(\min(X, Y) \geq z) = P(X \geq z, Y \geq z) = P(X \geq z)\, P(Y \geq z),$$

so by Example 11

$$P(\min(X, Y) \geq z) = (1 - p)^z(1 - p)^z = (1 - p)^{2z}.$$

It follows from Example 11 that $\min(X, Y)$ has a geometric distribution with parameter

$$1 - (1 - p)^2 = 2p - p^2.$$

To solve (b) we observe that

$$\begin{aligned}
P(Y \geq X) &= \sum_{x=0}^{\infty} P(X = x, Y \geq X) \\
&= \sum_{x=0}^{\infty} P(X = x, Y \geq x) \\
&= \sum_{x=0}^{\infty} P(X = x)P(Y \geq x) \\
&= \sum_{x=0}^{\infty} p(1 - p)^x(1 - p)^x \\
&= p \sum_{x=0}^{\infty} (1 - p)^{2x} \\
&= p/(1 - (1 - p)^2) = p/(2p - p^2).
\end{aligned}$$

To solve (c) we let z be a nonnegative integer. Then

$$\begin{aligned}
P(X + Y = z) &= \sum_{x=0}^{z} P(X = x, X + Y = z) \\
&= \sum_{x=0}^{z} P(X = x, Y = z - x) \\
&= \sum_{x=0}^{z} P(X = x)P(Y = z - x) \\
&= \sum_{x=0}^{z} p(1 - p)^x p(1 - p)^{z-x} \\
&= (z + 1)p^2(1 - p)^z.
\end{aligned}$$

The solution to (d) is given by

$$\begin{aligned}
P(Y = y \mid X + Y = z) &= \frac{P(Y = y, X + Y = z)}{P(X + Y = z)} \\
&= \frac{P(X = z - y, Y = y)}{P(X + Y = z)} \\
&= \frac{P(X = z - y)P(Y = y)}{P(X + Y = z)} \\
&= \frac{p(1 - p)^{z-y}p(1 - p)^y}{(z + 1)p^2(1 - p)^z} \\
&= \frac{1}{z + 1}.
\end{aligned}$$

Consider some experiment (such as rolling a die) that has only a finite or countably infinite number of possible outcomes. Then, as already explained, we can think of this experiment as that of observing the value of a discrete random variable X. Suppose the experiment is repeated n times. The combined experiment can be described as that of observing the values of the random variables $X_1, X_2, \ldots, X_n$, where X_i is the outcome of the ith experiment. If the experiments are repeated under identical conditions, presumably the chance mechanism remains the same, so we should require that these n random variables all have the same density. The intuitive notion that the repeated experiments have no influence on each other can now be formulated by demanding that the random variables $X_1, X_2, \ldots, X_n$ be mutually independent. Thus, in summary, n independent random variables $X_1, \ldots, X_n$ having a common discrete density f can be used to represent an n-fold independent repetition of an experiment having a finite or countably infinite number of outcomes.

The simplest random experiments are those that have only two possible outcomes, which we may label as success and failure. In tossing a coin, for example, we may think of getting a head as a success, while in drawing a card from a deck of r cards we may consider getting an ace as a success. Suppose we make n independent repetitions of our simple experiment. We can then describe the situation by letting $X_1, X_2, \ldots, X_n$ be n independent indicator random variables such that $X_i = 1$ or 0 according as the ith trial of the experiment results in a success or failure. In the literature, trials that can result in either success or failure are called *Bernoulli trials*, and the above situation is described by saying we perform n Bernoulli trials with common probability $p = P(X_i = 1)$ for success. In this context a random variable that takes on the values 1 and 0 with probabilities p and $1 - p$ respectively is said to have a *Bernoulli density* with parameter p.

The outcome of performing n Bernoulli trials can be given by the random vector $\mathbf{X} = (X_1, X_2, \ldots, X_n)$. The information conveyed in this vector tells exactly which trials were a success and which were a failure. Often, such precise information is not required, and all we want to know is the number S_n of trials that yielded a success among the n trials. In Example 2 we showed that S_n was binomially distributed with parameters n and p. Observe that $S_n = X_1 + \cdots + X_n$. Any random variable Y that is binomially distributed with these same parameters can be thought of as the sum of n independent Bernoulli random variables $X_1, \ldots, X_n$ each having parameter p.

Let us now consider independent repetitions of an experiment that has a finite number $r \geq 2$ of possible outcomes.

3.4.1. The multinomial distribution. Consider an experiment, such as rolling a die, that can result in only a finite number r of distinct possible outcomes. We may represent this experiment by saying we observe a random variable Y that assumes the values $1, 2, \ldots, r$, so that the event $\{Y = i\}$ represents the fact that the experiment yielded the ith outcome. Let $p_i = P(Y = i)$. If we perform an n-fold independent repetition of the experiment, we can represent the outcome of these n trials as an n-dimensional random vector $(Y_1, \ldots, Y_n)$, where the random variable Y_j corresponds to the jth trial. Here the random variables $Y_1, \ldots, Y_n$ are mutually independent and $P(Y_j = i) = p_i$.

The random vector $(Y_1, \ldots, Y_n)$ tells us the outcomes of these n trials. As in the case of $r = 2$ outcomes we often are not interested in such a detailed account, but only would like to know how many of the n trials resulted in each of the various possible outcomes. Let X_i, $i = 1, 2, \ldots, r$, denote the number of trials that yield the ith outcome. Then $X_i = x_i$ if and only if exactly x_i of the n random variables $Y_1, \ldots, Y_n$ assume the value i, i.e., exactly x_i of the n trials yield the ith outcome.

For example, for $r = 3$, $n = 5$, if

$$Y_1 = 2,\ Y_2 = 3,\ Y_3 = 3,\ Y_4 = 2, \text{ and } Y_5 = 2,$$

then

$$X_1 = 0,\ X_2 = 3, \text{ and } X_3 = 2.$$

We will now compute the joint density of $X_1, \ldots, X_r$. To this end let $x_1, x_2, \ldots, x_r$ be r nonnegative integers with sum $x_1 + \cdots + x_r = n$. A moment's thought shows that since the random variables $Y_1, Y_2, \ldots, Y_n$ are independent with a common density, every specified choice of x_1 of them having value 1, x_2 of them having value $2, \ldots, x_r$ of them having value r, has the same probability, namely

$$p_1^{x_1} p_2^{x_2} \cdots p_r^{x_r}.$$

Thus letting $C(n; x_1, \ldots, x_r)$ denote the number of possible choices, we see that

$$P(X_1 = x_1, \ldots, X_r = x_r) = C(n; x_1, \ldots, x_r)\, p_1^{x_1} \cdots p_r^{x_r}.$$

The computation of $C(n; x_1, \ldots, x_r)$ is a problem in combinatorial analysis that can easily be solved by the methods of Chapter 2. The simplest way of doing this is to think of the r values $1, 2, \ldots, r$ as r boxes and the n trials as n balls. Then $C(n; x_1, \ldots, x_r)$ is the number of ways we can place the n balls into the r boxes in such a manner as to have exactly x_1 balls in box $1, \ldots$, exactly x_r balls in box r. If this is so, then box 1 has x_1 balls. These x_1 balls can be chosen from the n balls in $\binom{n}{x_1}$ ways. The remaining $n - x_1$ balls must be placed into the $r - 1$ boxes $2, \ldots, r$ in such a manner as to have x_2 balls in box $2, \ldots, x_r$ balls in box r. Thus

$$(16) \qquad C(n; x_1, \ldots, x_r) = \binom{n}{x_1} C(n - x_1; x_2, \ldots, x_r).$$

It now follows by induction on r that

$$(17) \qquad C(n; x_1, \ldots, x_r) = \frac{n!}{(x_1!)(x_2!)\cdots(x_r!)}.$$

Indeed, for $r = 1$ there is nothing to prove. Assume that (17) holds for $r - 1$ boxes. Then from (16) we see that

$$C(n; x_1, \ldots, x_r) = \frac{n!}{(x_1!)(n - x_1)!} \frac{(n - x_1)!}{(x_2!)\cdots(x_r!)}$$

$$= \frac{n!}{(x_1!)\cdots(x_r!)}$$

as desired.

The joint density f of $X_1, \ldots, X_r$ is therefore given by

$$(18) \quad f(x_1, \ldots, x_r) = \begin{cases} \dfrac{n!}{(x_1!)\cdots(x_r!)} p_1^{x_1} \cdots p_r^{x_r}, \\ \qquad x_i \text{ integers} \geq 0 \text{ such that } x_1 + \cdots + x_r = n, \\ 0, \quad \text{elsewhere.} \end{cases}$$

This density is called the *multinomial density* with parameters n and $p_1, \ldots, p_r$.

We observe at once that the r random variables $X_1, \ldots, X_r$ are *not* independent. In fact, since $X_1 + \cdots + X_r = n$, any $r - 1$ of them determine the rth. This, plus the fact that $p_1 + \cdots + p_r = 1$, is sometimes used to express the multinomial distribution in a different form. Let $x_1, x_2, \ldots, x_{r-1}$ be $r - 1$ nonnegative integers such that $x_1 + \cdots + x_{r-1} \leq n$. Then

$$(19) \quad P(X_1 = x_1, \ldots, X_{r-1} = x_{r-1})$$
$$= \frac{n!}{(x_1!)\cdots(x_{r-1}!)(n - x_1 - \cdots - x_{r-1})!}$$
$$\times p_1^{x_1} \cdots p_{r-1}^{x_{r-1}}(1 - p_1 - \cdots - p_{r-1})^{n - x_1 - \cdots - x_{r-1}}.$$

This form is convenient when we are interested in the first $r - 1$ outcomes and think of the rth outcome as the outcome which is "not one of the $r - 1$ outcomes." Thus in rolling a die we might be interested in only knowing if a 2, 4, or 6 appeared. The experiment would then have four possible outcomes "2," "4," "6," and "not (2, 4, 6)."

Let k be a nonnegative integer, $k \leq r$. A simple probability argument shows that for $x_1, x_2, \ldots, x_k$ nonnegative integers such that $x_1 + \cdots + x_k \leq n$,

$$(20) \quad P(X_1 = x_1, \ldots, X_k = x_k)$$
$$= \frac{n!}{(x_1!)\cdots(x_k!)[n - (x_1 + \cdots + x_k)]!}$$
$$\times p_1^{x_1} \cdots p_k^{x_k}(1 - (p_1 + \cdots + p_k))^{n - (x_1 + \cdots + x_k)}.$$

To see this, observe that in performing the n trials we are now only interested in the $k + 1$ outcomes "1," "2," . . . , "k," and "not $(1, 2, \ldots, k)$." Thus in essence we have n repeated trials of an experiment having $k + 1$ outcomes, with X_i being the number of times that the ith outcome occurs, $i = 1, 2, \ldots, k$. Equation (20) now follows from (19) with $r - 1 = k$.

3.4.2. Poisson approximation to the binomial distribution. There is an important connection between the binomial distribution and the Poisson distribution. Suppose, for example, that we perform n Bernoulli trials with success probability $p_n = \lambda/n$ at each trial. Then the probability of having $S_n = k$ successes in the n trials is given by

$$P(S_n = k) = \binom{n}{k} (p_n)^k (1 - p_n)^{n-k}$$
$$= \frac{\lambda^k}{k!} \frac{(n)_k}{n^k} \left(1 - \frac{\lambda}{n}\right)^n \left(1 - \frac{\lambda}{n}\right)^{-k}.$$

Now as $n \to \infty$, $(n)_k/n^k \to 1$, $(1 - \lambda/n)^n \to e^{-\lambda}$, and $(1 - \lambda/n)^{-k} \to 1$. Consequently,

$$\lim_{n\to\infty} \binom{n}{k} (p_n)^k (1 - p_n)^{n-k} = \frac{\lambda^k}{k!} e^{-\lambda}. \tag{21}$$

In the derivation of (21) we had $np_n = \lambda$. Actually (21) holds whenever $np_n \to \lambda$ as $n \to \infty$.

Equation (21) is used in applications to approximate the binomial distribution by the Poisson distribution when the success probability p is small and n is large. This is done by approximating the binomial probability $P(S_n = x)$ by means of $f(x)$ where f is the Poisson density with parameter $\lambda = np$. The approximation is quite good if np^2 is small. The following example illustrates the use of this technique.

Example 15. A machine produces screws, 1% of which are defective. Find the probability that in a box of 200 screws there are no defectives.

Here we have $n = 200$ Bernoulli trials, with success probability $p = .01$. The probability that there are no defective screws is

$$(1 - .01)^{200} = (.99)^{200} = .1340.$$

The Poisson approximation to this is given by

$$e^{-200(.01)} = e^{-2} = .1353.$$

The fact that the Poisson distribution can arise as a limit of binomial distributions has important theoretical consequences. It is one justification for developing models based on Poisson processes, which will be discussed in Chapter 9. The use of the Poisson approximation as a labor-saving device in computing binomial probabilities is of secondary importance, since the binomial probabilities themselves are readily computed.

3.5. Infinite sequences of Bernoulli trials

Consider repeatedly performing a success-failure experiment having probability p of success until the first success appears. For any prescribed number n of trials, there is the nonzero probability $(1 - p)^n$ that no success occurs. Thus, in considering the number of trials until the first success, we cannot limit ourselves to any prescribed number of trials, but instead must consider an unending sequence of trials.

A given finite number n of trials constitute n Bernoulli trials, represented by n independent Bernoulli random variables, $X_1, \ldots, X_n$. To represent an infinite sequence of Bernoulli trials we consider an infinite sequence $\{X_n\}$, $n \geq 1$, of independent Bernoulli random variables having the same parameter p.

In general, random variables $X_1, X_2, \ldots$ are said to be independent if for any positive integer n, the random variables $X_1, \ldots, X_n$ are mutually independent. It can be shown that, given any discrete density f, there is a probability space $(\Omega, \mathscr{A}, P)$ upon which are defined mutually independent random variables $X_1, X_2, \ldots$ each having the density f.

As our model for performing an unlimited sequence of Bernoulli trials, we therefore take an infinite sequence $\{X_n\}$, $n \geq 1$, of mutually independent Bernoulli random variables such that $P(X_n = 1) = p$, $n \geq 1$. We interpret $X_n = 1$ as meaning that the nth trial results in success, and $X_n = 0$ as meaning that it results in failure.

Consider the number of trials W_1 until the first success. The random variable W_1 can assume only the integer values $1, 2, \ldots$. The event $\{W_1 = n\}$ occurs if and only if the first $n - 1$ trials are failures and the nth trial is a success. Therefore

$$\{W_1 = n\} = \{X_1 = 0, \ldots, X_{n-1} = 0, X_n = 1\}.$$

It follows that

$$\begin{aligned} P(W_1 = n) &= P(X_1 = 0, \ldots, X_{n-1} = 0, X_n = 1) \\ &= P(X_1 = 0) \cdots P(X_{n-1} = 0)P(X_n = 1) \\ &= (1 - p)^{n-1}p. \end{aligned}$$

Consequently

$$P(W_1 - 1 = n) = p(1 - p)^n. \tag{22}$$

Thus $W_1 - 1$ is geometrically distributed with parameter p.

Let $r \geq 1$ be an integer and let T_r denote the number of trials until the rth success (so that the rth success occurs at trial T_r). Then T_r is a random variable that can assume only the integer values $r, r + 1, \ldots$. The event $\{T_r = n\}$ occurs if and only if there is a success at the nth trial and during the first $n - 1$ trials there are exactly $r - 1$ successes. Thus

$$\{T_r = n\} = \{X_1 + \cdots + X_{n-1} = r - 1\} \cap \{X_n = 1\}.$$

Since the two events on the right are independent and $X_1 + \cdots + X_{n-1}$ has a binomial distribution with parameters $n - 1$ and p, we see that for $n = r, r + 1, \ldots$

$$\begin{aligned} P(T_r = n) &= P(X_1 + \cdots + X_{n-1} = r - 1)P(X_n = 1) \\ &= \binom{n-1}{r-1} p^{r-1}(1 - p)^{n-r}p \\ &= \binom{n-1}{r-1} p^r(1 - p)^{n-r}. \end{aligned}$$

Consequently

$$P(T_r - r = n) = \binom{r + n - 1}{r - 1} p^r(1 - p)^n. \tag{23}$$

We see from Equations (4) and (23) that $T_r - r$ has the negative binomial distribution with parameters $\alpha = r$ and p.

Let $T_0 = 0$ and for any integer $r \geq 1$ let T_r be as above. Define $W_i = T_i - T_{i-1}$, $i = 1, 2, \ldots$. Then W_i is the number of trials after the $(i - 1)$st success until the ith success. We will now show that for any integer $r \geq 1$ the random variables $W_1 - 1, W_2 - 1, \ldots, W_r - 1$ are mutually independent and have the same geometric density with parameter p.

To see this let $n_1, n_2, \ldots, n_r$ be any r positive integers. Then the event $\{W_1 = n_1, \ldots, W_r = n_r\}$ occurs if and only if among the first $n_1 + \cdots + n_r$ trials all are failures except for trials

$$n_1, n_1 + n_2, \ldots, n_1 + \cdots + n_r,$$

which are successes. Since the trials are mutually independent with success probability p we see that

$$\begin{aligned} P(W_1 = n_1, \ldots, W_r = n_r) &= (1 - p)^{n_1 - 1}p(1 - p)^{n_2 - 1}p \cdots (1 - p)^{n_r - 1}p \\ &= \prod_{i=1}^{r} [p(1 - p)^{n_i - 1}]. \end{aligned}$$

Thus the random variables $W_1 - 1, \ldots, W_r - 1$ are independent, geometrically distributed with parameter p.

Now clearly $T_r - r = (W_1 - 1) + \cdots + (W_r - 1)$, so we see that $T_r - r$ is the sum of r independent, geometrically distributed random variables. We have previously found that $T_r - r$ is negative binomially distributed with parameters r and p. We have thus established the interesting and important fact that *the distribution of the sum of r independent, identically distributed geometric random variables with parameter p is negative binomially distributed with parameters r and p.*

Further properties of infinite sequences of independent Bernoulli trials will be treated in the exercises.

3.6. Sums of independent random variables

In this section we discuss methods for finding the distribution of the sum of a finite number of independent discrete random variables. Let us start by considering two such variables X and Y.

We assume, then, that X and Y are independent discrete random variables. Let $x_1, x_2, \ldots$ denote the distinct possible values of X. For any z, the event $\{X + Y = z\}$ is the same as the event

$$\bigcup_i \{X = x_i, Y = z - x_i\}.$$

Since the events $\{X = x_i, Y = z - x_i\}$ are disjoint for distinct values of i, it follows that

$$\begin{aligned} P(X + Y = z) &= \sum_i P(X = x_i, Y = z - x_i) \\ &= \sum_i P(X = x_i)P(Y = z - x_i) \\ &= \sum_i f_X(x_i)f_Y(z - x_i). \end{aligned}$$

In other words

$$f_{X+Y}(z) = \sum_x f_X(x)f_Y(z - x). \tag{24}$$

If X and Y are integer-valued random variables, then $X + Y$ is also integer-valued. In this case we can interpret (24) as being valid when z is an integer and the variable x in the right-hand side of (24) ranges over the integers. One further specialization is useful. Suppose that X and Y assume only nonnegative integer values. Then $X + Y$ also assumes only nonnegative integer values. If z is a nonnegative integer, then $f_X(x)f_Y(z - x) = 0$ unless x is one of the values $0, 1, \ldots, z$. Thus under these assumptions (24) can be written as

$$f_{X+Y}(z) = \sum_{x=0}^{z} f_X(x)f_Y(z - x). \tag{25}$$

Although Equation (25) is useful for computing the density of $X + Y$, it is usually simpler to use probability generating functions. We will next describe such functions and then give several important applications of their use in computing the density of the sum of independent random variables.

Definition 6 *Let X be a nonnegative integer-valued random variable. The probability generating function Φ_X of X is defined as*

$$\Phi_X(t) = \sum_{x=0}^{\infty} P(X = x)t^x = \sum_{x=0}^{\infty} f_X(x)t^x, \qquad -1 \le t \le 1.$$

We will now calculate $\Phi_X(t)$ in three specific cases.

Example 16. Binomial distribution. Let X have a binomial distribution with parameters n and p. Then

$$P(X = x) = \binom{n}{x} p^x(1 - p)^{n-x}$$

and hence

$$\Phi_X(t) = \sum_{x=0}^{n} P(X = x)t^x = \sum_{x=0}^{n} \binom{n}{x} (pt)^x(1 - p)^{n-x}.$$

From the binomial expansion formula

$$(a + b)^n = \sum_{x=0}^{n} \binom{n}{x} a^x b^{n-x},$$

we conclude that

$$\Phi_X(t) = (pt + 1 - p)^n. \tag{26}$$

Example 17. Negative binomial distribution. Let X have a negative binomial distribution with parameters α and p. Then

$$P(X = x) = p^\alpha \binom{-\alpha}{x} (-1)^x(1 - p)^x$$

and hence

$$\begin{aligned}\Phi_X(t) &= \sum_{x=0}^{\infty} p^\alpha \binom{-\alpha}{x} (-1)^x(1 - p)^x t^x \\ &= p^\alpha \sum_{x=0}^{\infty} \binom{-\alpha}{x} (-t(1 - p))^x.\end{aligned}$$

From the Taylor series expansion

$$(1 + s)^{-\alpha} = \sum_{x=0}^{\infty} \binom{-\alpha}{x} s^x,$$

with $s = -t(1 - p)$, it follows that

$$\Phi_X(t) = \left(\frac{p}{1 - t(1 - p)}\right)^\alpha. \tag{27}$$

Example 18. Poisson distribution. Let X have a Poisson distribution with parameter λ. Then

$$P(X = x) = \frac{\lambda^x e^{-\lambda}}{x!}$$

and hence

$$\Phi_X(t) = e^{-\lambda} \sum_{x=0}^{\infty} \frac{(\lambda t)^x}{x!}.$$

By setting $s = \lambda t$ in the Taylor series expansion

$$e^s = \sum_{x=0}^{\infty} \frac{s^x}{x!},$$

we see that

$$\Phi_X(t) = e^{\lambda t} e^{-\lambda} = e^{\lambda(t-1)}. \tag{28}$$

Let X and Y be independent, nonnegative integer-valued random variables. Then

$$\Phi_{X+Y}(t) = \Phi_X(t)\Phi_Y(t). \tag{29}$$

To see this, note that by (25)

$$\begin{aligned}
\Phi_{X+Y}(t) &= \sum_{z=0}^{\infty} f_Z(z)t^z \\
&= \sum_{z=0}^{\infty} t^z \sum_{x=0}^{z} f_X(x)f_Y(z-x) \\
&= \sum_{x=0}^{\infty} f_X(x)t^x \sum_{z=x}^{\infty} f_Y(z-x)t^{z-x} \\
&= \sum_{x=0}^{\infty} f_X(x)t^x \sum_{y=0}^{\infty} f_Y(y)t^y \\
&= \Phi_X(t)\Phi_Y(t),
\end{aligned}$$

which is the desired result.

It follows easily from (29) by induction that if $X_1, \ldots, X_r$ are independent, nonnegative integer-valued random variables, then

$$\Phi_{X_1+\cdots+X_r}(t) = \Phi_{X_1}(t)\cdots\Phi_{X_r}(t). \tag{30}$$

The conclusions of the next theorem can be proven most easily by the "generating function technique," which is based upon the fact that if

$$\sum_{x=0}^{\infty} a_x t^x = \sum_{x=0}^{\infty} b_x t^x, \qquad -1 < t < 1,$$

then we may equate the coefficients of t^x in the two power series and conclude that $a_x = b_x$, $x = 0, 1, 2, \ldots$. This shows that if two nonnegative integer-valued random variables have the same probability generating function, they must have the same distribution. In other words, the probability generating function of a nonnegative integer-valued random variable uniquely determines the distribution of that random variable.

Theorem 1 *Let $X_1, \ldots, X_r$ be independent random variables.*

(i) *If X_i has the binomial distribution with parameters n_i and p, then $X_1 + \cdots + X_r$ has the binomial distribution with parameters $n_1 + \cdots + n_r$ and p.*

(ii) *If X_i has the negative binomial distribution with parameters α_i and p, then $X_1 + \cdots + X_r$ has the negative binomial distribution with parameters $\alpha_1 + \cdots + \alpha_r$ and p.*

(iii) *If X_i has the Poisson distribution with parameter λ_i, then $X_1 + \cdots + X_r$ has the Poisson distribution with parameter $\lambda_1 + \cdots + \lambda_r$.*

Proof of (i). If the X_i's are as in (i), then by Example 16

$$\begin{aligned}\Phi_{X_1 + \cdots + X_r}(t) &= \Phi_{X_1}(t) \cdots \Phi_{X_r}(t) \\ &= (pt + 1 - p)^{n_1} \cdots (pt + 1 - p)^{n_r} \\ &= (pt + 1 - p)^{n_1 + \cdots + n_r}.\end{aligned}$$

Thus the probability generating function of $X_1 + \cdots + X_r$ is the same as that of a random variable having a binomial distribution with parameters $n_1 + \cdots + n_r$ and p. This implies that $X_1 + \cdots + X_r$ must have that binomial distribution. For let

$$a_x = \binom{n_1 + \cdots + n_r}{x} p^x (1 - p)^{n_1 + \cdots + n_r - x}$$

denote the corresponding binomial probabilities. Then

$$\begin{aligned}\sum_{x=0}^{\infty} P(X_1 + \cdots + X_r = x) t^x &= \Phi_{X_1 + \cdots + X_r}(t) \\ &= (pt + 1 - p)^{n_1 + \cdots + n_r} \\ &= \sum_{x=0}^{\infty} a_x t^x.\end{aligned}$$

Thus by equating coefficients we see that

$$P(X_1 + \cdots + X_r = x) = a_x$$

and hence that $X_1 + \cdots + X_r$ is binomially distributed as stated in (i).

Proof of (ii). If the X_i's are as in (ii), then by Example 17

$$\Phi_{X_1 + \cdots + X_r}(t) = \Phi_{X_1}(t) \cdots \Phi_{X_r}(t)$$

$$= \left(\frac{p}{1 - t(1 - p)}\right)^{\alpha_1} \cdots \left(\frac{p}{1 - t(1 - p)}\right)^{\alpha_r}$$

$$= \left(\frac{p}{1 - t(1 - p)}\right)^{\alpha_1 + \cdots + \alpha_r}.$$

Thus the probability generating function of $X_1 + \cdots + X_r$ is the same as that of a random variable having a negative binomial distribution with parameters $\alpha_1 + \cdots + \alpha_r$ and p. It now follows by the same argument used in proving (i) that $X_1 + \cdots + X_r$ has that negative binomial distribution.

The proof of (iii) is similar to that of (i) and (ii) and is left as an exercise for the reader. ∎

Suppose $\alpha_1 = \cdots = \alpha_r = 1$ in statement (ii) of Theorem 1. Then $X_1, \ldots, X_r$ are each geometrically distributed with parameter p, and (ii) states that $X_1 + \cdots + X_r$ has a negative binomial distribution with parameters r and p. This provides an alternative proof to the result obtained in Section 3.5.

The next example illustrates the use of conditional probabilities.

Example 19 Let $X_1, X_2, \ldots$ be independent nonnegative integer-valued random variables having a common density. Set $S_0 = 0$ and $S_n = X_1 + \cdots + X_n$, $n \geq 1$. Let N be a nonnegative integer-valued random variable and suppose that $N, X_1, X_2, \ldots$ are independent. Then $S_N = X_1 + \cdots + X_N$ is the sum of a random number of random variables. For an interpretation of S_N suppose that at time 0 a random number N of bacteria enters a system and that by time 1 the colony started by the ith bacterium contains X_i members. Then S_N is the total number of bacteria present at time 1. Show that the probability generating function of S_N is given by

$$\Phi_{S_N}(t) = \Phi_N(\Phi_{X_1}(t)), \qquad -1 \leq t \leq 1. \tag{31}$$

To verify (31) we observe first that

$$\begin{aligned} P(S_N = x) &= \sum_{n=0}^{\infty} P(S_N = x, N = n) \\ &= \sum_{n=0}^{\infty} P(S_n = x, N = n) \\ &= \sum_{n=0}^{\infty} P(N = n) P(S_n = x \mid N = n). \end{aligned}$$

Since N is independent of $X_1, X_2, \ldots, X_n$, it is independent of S_n, and hence $P(S_n = x \mid N = n) = P(S_n = x)$. Thus

$$P(S_N = x) = \sum_{n=0}^{\infty} P(N = n) P(S_n = x). \tag{32}$$

Consequently for $-1 \leq t \leq 1$

$$\begin{aligned}
\Phi_{S_N}(t) &= \sum_{x=0}^{\infty} t^x P(S_N = x) \\
&= \sum_{x=0}^{\infty} t^x \sum_{n=0}^{\infty} P(N = n)P(S_n = x) \\
&= \sum_{n=0}^{\infty} P(N = n) \sum_{x=0}^{\infty} t^x P(S_n = x) \\
&= \sum_{n=0}^{\infty} P(N = n)\, \Phi_{S_n}(t) \\
&= \sum_{n=0}^{\infty} P(N = n)(\Phi_{X_1}(t))^n = \Phi_N(\Phi_{X_1}(t)).
\end{aligned}$$

Exercises

1 Any point in the interval $[0, 1)$ can be represented by its decimal expansion $.x_1x_2\ldots$. Suppose a point is chosen at random from the interval $[0, 1)$. Let X be the first digit in the decimal expansion representing the point. Compute the density of X.

2 Let X have the negative binomial density with parameters $\alpha = r$ (r an integer) and p. Compute the density of $X + r$.

3 Suppose a box has 6 red balls and 4 black balls. A random sample of size n is selected. Let X denote the number of red balls selected. Compute the density of X if the sampling is (a) without replacement, (b) with replacement.

4 Let N be a positive integer and let

$$f(x) = \begin{cases} c2^x, & x = 1, 2, \ldots, N, \\ 0, & \text{elsewhere.} \end{cases}$$

Find the value of c such that f is a probability density.

5 Suppose X is a random variable having density f given by

x	-3	-1	0	1	2	3	5	8
$f(x)$	.1	.2	.15	.2	.1	.15	.05	.05

Compute the following probabilities:
(a) X is negative;
(b) X is even;
(c) X takes a value between 1 and 8 inclusive;
(d) $P(X = -3 \mid X \leq 0)$;
(e) $P(X \geq 3 \mid X > 0)$.

6 Suppose X has a geometric distribution with $p = .8$. Compute the probabilities of the following events:
(a) $X > 3$;
(b) $4 \leq X \leq 7$ or $X > 9$;
(c) $3 \leq X \leq 5$ or $7 \leq X \leq 10$.

7 Let X be uniformly distributed on $0, 1, \ldots, 99$. Calculate
(a) $P(X \geq 25)$;
(b) $P(2.6 < X < 12.2)$;
(c) $P(8 < X \leq 10 \text{ or } 30 < X \leq 32)$;
(d) $P(25 \leq X \leq 30)$.

8 Suppose a box has 12 balls labeled $1, 2, \ldots, 12$. Two independent repetitions are made of the experiment of selecting a ball at random from the box. Let X denote the larger of the two numbers on the balls selected. Compute the density of X.

9 Suppose the situation is as in Exercise 8, except now the two balls are selected without replacement. Compute the density of X.

10 Let X be a geometrically distributed random variable having parameter p. Let $Y = X$ if $X < M$ and let $Y = M$ if $X \geq M$; that is, $Y = \text{Min}(X, M)$. Compute the density of Y.

11 Let X be geometrically distributed with parameter p. Compute the density of
(a) X^2;
(b) $X + 3$.

12 Suppose a box has r balls numbered $1, 2, \ldots, r$. A random sample of size n is selected without replacement. Let Y denote the largest of the numbers drawn and let Z denote the smallest.
(a) Compute the probability $P(Y \leq y)$.
(b) Compute the probability $P(Z \geq z)$.

13 Let X and Y be two random variables having the joint density given by the following table.

$X \backslash Y$	-1	0	2	6
-2	1/9	1/27	1/27	1/9
1	2/9	0	1/9	1/9
3	0	0	1/9	4/27

Compute the probability of the following events:
(a) Y is even;
(b) XY is odd;
(c) $X > 0$ and $Y \geq 0$.

14 Let X and Y be independent random variables each having the uniform density on $\{0, 1, \ldots, N\}$. Find
(a) $P(X \geq Y)$;
(b) $P(X = Y)$.

15 Let X and Y be as in Exercise 14. Find the densities of
(a) $\min(X, Y)$;
(b) $\max(X, Y)$;
(c) $|Y - X|$.

add to HW assignment

16 Let X and Y be independent random variables having geometric densities with parameters p_1 and p_2 respectively. Find
(a) $P(X \geq Y)$;
(b) $P(X = Y)$.

17 Let X and Y be as in Exercise 16. Find the density of
(a) $\min(X, Y)$;
(b) $X + Y$.

18 Let X and Y be discrete random variables and let g and h be functions such that the following identity holds:

$$P(X = x, Y = y) = g(x)h(y).$$

(a) Express $P(X = x)$ in terms of g and h.
(b) Express $P(Y = y)$ in terms of g and h.
(c) Show that $(\sum_x g(x))(\sum_y h(y)) = 1$.
(d) Show that X and Y are independent.

19 Let X and Y be independent random variables each having a geometric density with parameter p. Set $Z = Y - X$ and $M = \min(X, Y)$.
(a) Show that for integers z and $m \geq 0$

$$P(M = m, Z = z) = \begin{cases} P(X = m - z)P(Y = m), & z < 0, \\ P(X = m)P(Y = m + z), & z \geq 0. \end{cases}$$

(b) Conclude from (a) that for integers z and $m \geq 0$

$$P(M = m, Z = z) = p^2(1 - p)^{2m}(1 - p)^{|z|}.$$

(c) Use (b) and Exercise 18 to show that M and Z are independent.

20 Suppose a circular target is divided into three zones bounded by concentric circles of radius 1/3, 1/2, and 1, as illustrated in the following diagram.

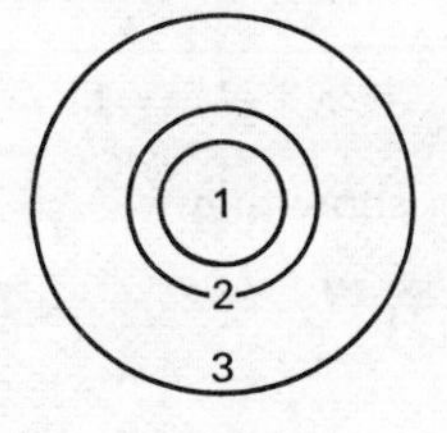

Figure 4

If three shots are fired at random at the target, what is the probability that exactly one shot lands in each zone?

21 Suppose $2r$ balls are distributed at random into r boxes. Let X_i denote the number of balls in box i.
(a) Find the joint density of $X_1, \ldots, X_r$.
(b) Find the probability that each box contains exactly 2 balls.

22 Consider an experiment having three possible outcomes that occur with probabilities p_1, p_2, and p_3, respectively. Suppose n independent repetitions of the experiment are made and let X_i denote the number of times the ith outcome occurs.

(a) What is the density of $X_1 + X_2$?

(b) Find $P(X_2 = y \mid X_1 + X_2 = z)$, $y = 0, 1, \ldots, z$.

23 Use the Poisson approximation to calculate the probability that at most 2 out of 50 given people will have invalid driver's licenses if normally 5% of the people do.

24 Use the Poisson approximation to calculate the probability that a box of 100 fuses has at most 2 defective fuses if 3% of the fuses made are defective.

25 A die is rolled until a 6 appears.

(a) What is the probability that at most six rolls are needed?

(b) How many rolls are required so that the probability of getting 6 is at least 1/2?

Exercises 26–30 are related problems concerning an infinite sequence of Bernoulli trials as discussed in Section 3.5.

26 Let T_i be the number of trials up to and including the ith success. Let $0 \leq x_1 < \cdots < x_r$ be integers. Compute the probability

$$P(T_1 = x_1, T_2 = x_2, \ldots, T_r = x_r).$$

Hint: Let $W_r = T_r - T_{r-1}$, $r \geq 2$, and let $W_1 = T_1$; then

$$P(T_1 = x_1, \ldots, T_r = x_r)$$
$$= P(W_1 = x_1, W_2 = x_2 - x_1, \ldots, W_r = x_r - x_{r-1}).$$

Now use the fact that the random variables $W_1 - 1, \ldots, W_r - 1$ are mutually independent random variables, each having a geometric distribution with parameter p.

27 Let N_n be the number of successes in the first n trials. Show that

$$P(T_1 = x \mid N_n = 1) = \frac{1}{n}, \qquad x = 1, 2, \ldots, n.$$

28 More generally, show that

$$P(T_1 = x_1, T_2 = x_2, \ldots, T_r = x_r \mid N_n = r) = \binom{n}{r}^{-1},$$
$$0 < x_1 < x_2 < \cdots < x_r \leq n.$$

This shows that given there are r successes during the first n trials, the trials at which these successes occur constitute a random sample of size r (without replacement) from the "population" of possible positions.

29 Let k be a positive integer, $k \leq r$. From Exercise 28 we may readily compute that

$$P(T_k = x \mid N_n = r) = \frac{\binom{x-1}{k-1}\binom{n-x}{r-k}}{\binom{n}{r}}.$$

Indeed, if $T_k = x$ then the kth success is at position x. In the first $x - 1$ positions there must be exactly $k - 1$ successes, and in the last $n - x$ positions there must be exactly $r - k$ successes. Since, given $N_n = r$, the positions of the r successes are a random sample of size r from the "population" of n positions, the result follows. Verify that this is in fact so, by computing directly $P(T_k = x \mid N_n = r)$.

30 Let $1 \le i < j \le r$ be nonnegative integers. Compute

$$P(T_i = x, T_j = y \mid N_n = r)$$

for $0 < x < y \le n$.

31 Suppose X and Y are independent random variables having the uniform density on $1, 2, \ldots, N$. Compute the density of $X + Y$.

32 Let X be uniformly distributed on $\{0, 1, 2, \ldots, N\}$. Find $\Phi_X(t)$.

33 Let X be a nonnegative integer-valued random variable whose probability generating function is given by $\Phi_X(t) = e^{\lambda(t^2-1)}$, where $\lambda > 0$. Find f_X.

34 Prove (iii) of Theorem 1.

35 Let X and Y be independent random variables having Poisson densities with parameters λ_1 and λ_2 respectively. Find $P(Y = y \mid X + Y = z)$ for $y = 0, \ldots, z$. *Hint:* Use (iii) of Theorem 1.

36 Let X, Y, and Z be independent random variables having Poisson densities with parameters λ_1, λ_2, and λ_3 respectively. Find

$$P(X = x, Y = y, Z = z \mid X + Y + Z = x + y + z)$$

for nonnegative integers x, y, and z. *Hint:* Use (iii) of Theorem 1.

37 In Example 19 suppose that X_1 takes on the values 1 and 0 with respective probabilities p and $1 - p$, where $0 < p < 1$. Suppose also that N has a Poisson density with parameter λ.

(a) Use Equation (31) to find the probability generating function of S_N.

(b) Use (a) to find the density of S_N.

For an interpretation of S_N suppose a random number N of cancer cells is introduced at time 0 and that each cell, independently of the other cells and independently of N, has probability p of surviving a treatment of radiation. Let $X_i = 1$ if the ith cell survives and let $X_i = 0$ otherwise. Then S_N is the number of cells that survive the treatment.

38 Solve (b) of Exercise 37 without using probability generating functions, but using instead Equation (32) and the fact that $X_1 + \cdots + X_n$ has a binomial density.

4 Expectation of Discrete Random Variables

Let us consider playing a certain game of chance. In order to play the game, we must pay a fee of a dollars. As a result of playing the game we receive X dollars, where X is a random variable having possible values $x_1, x_2, \ldots, x_r$. The question is, should we play the game? If the game is to be played only once, then this question is quite difficult. However, suppose we play the game a large number of times. After n plays we would pay na dollars and receive $X_1 + \cdots + X_n$ dollars. If we assume that the successive plays of the game constitute independent repetitions of the same experiment (observing a value of X), then we can take the random variables $X_1, X_2, \ldots, X_n$ as mutually independent and having the common density f of X. Let $N_n(x_i)$ denote the number of games that yielded the value x_i, i.e., the number of X_i's that assume the value x_i. Then we can write

$$X_1 + \cdots + X_n = \sum_{i=1}^{r} x_i N_n(x_i).$$

The average amount received is then

$$\frac{X_1 + \cdots + X_n}{n} = \sum_{i=1}^{r} x_i \left[\frac{N_n(x_i)}{n}\right].$$

According to the relative frequency interpretation of probabilities, if n is large, the numbers $N_n(x_i)/n$ should be approximately equal to $f(x_i)$, and thus the sum on the right should be approximately equal to $\mu = \sum_{i=1}^{r} x_i f(x_i)$. Thus it seems reasonable to anticipate a net gain in playing the game if $\mu > a$ and to expect a net loss if $\mu < a$. If $\mu = a$ we would anticipate just about breaking even.

The quantity $\sum_{i=1}^{r} x_i f(x_i)$ is called the expectation of the random variable X. More generally, let X be any discrete random variable that assumes a finite number of values $x_1, \ldots, x_r$. Then the expected value of X, denoted by EX or μ, is the number

$$EX = \sum_{i=1}^{r} x_i f(x_i), \tag{1}$$

where f is the density of X.

Suppose X has the uniform distribution on the set $\{x_1, \ldots, x_r\}$. Then $f(x_i) = P(X = x_i) = r^{-1}$, and from (1) we see that $EX = (x_1 + \cdots + x_r)r^{-1}$, so in this case EX is just the arithmetic average of the possible values of X. In general, (1) shows that EX is a *weighted average* of the possible values of X; the weight attached to the ith value x_i is its probability $f(x_i)$.

The expected value EX is also called the mean of X (or of the density f of X) and is customarily denoted by μ. The mean is one way of trying to summarize a probability distribution by a single number that is supposed to represent a "typical value" of X. How good this is depends on how closely the values of X are clustered about the value μ. We will examine this question in more detail when we discuss the variance of X in Section 4.3.

Example 1. Binomial distribution. Let X have the binomial distribution with parameters n and p. Find EX.

For $n = 1$, X assumes the two values 0 and 1 with probabilities $(1 - p)$ and p respectively. Hence

$$EX = 0 \cdot P(X = 0) + 1 \cdot P(X = 1) = p.$$

Since a random variable having a binomial density with parameters 1 and p is just an indicator random variable, we see that we can find the probability of the event A that $X = 1$ by computing the expectation of its indicator.

We now compute EX for any $n \geq 1$. In this case X assumes the values $0, 1, 2, \ldots, n$, and

$$EX = \sum_{j=0}^{n} j \binom{n}{j} p^j (1 - p)^{n-j}.$$

To calculate this quantity we observe that

$$\begin{aligned} j \binom{n}{j} &= \frac{jn!}{j!\,(n - j)!} \\ &= \frac{n(n - 1)!}{(j - 1)!\,[(n - 1) - (j - 1)]!} \\ &= n \binom{n - 1}{j - 1}. \end{aligned}$$

Thus

$$EX = n \sum_{j=1}^{n} \binom{n - 1}{j - 1} p^j (1 - p)^{n-j}.$$

Making the change of variable $i = j - 1$ we see that

$$EX = np \sum_{i=0}^{n-1} \binom{n - 1}{i} p^i (1 - p)^{n-i-1}.$$

By the binomial theorem

$$\sum_{i=0}^{n-1} \binom{n - 1}{i} p^i (1 - p)^{n-i-1} = [p + (1 - p)]^{n-1} = 1,$$

so we see that

$$EX = np.$$

4.1. Definition of expectation

Suppose now that X is any discrete random variable having possible values $x_1, x_2, \ldots$. We would like to define the expectation of X as

$$EX = \sum_{j=1}^{\infty} x_j f(x_j). \tag{2}$$

If X has only a finite number of possible values $x_1, \ldots, x_r$, then (2) is just our previous definition. In the general discrete case, this definition is valid provided that the sum $\sum_j x_j f(x_j)$ is well defined. For this to be the case we require that $\sum_j |x_j| f(x_j) < \infty$. This leads to the following.

Definition 1 *Let X be a discrete random variable having density f. If $\sum_j |x_j| f(x_j) < \infty$, then we say that X has finite expectation and we define its expectation by (2). On the other hand if $\sum_{j=1}^{\infty} |x_j| f(x_j) = \infty$, then we say X does not have finite expectation and EX is undefined.*

If X is a nonnegative random variable, the fact that X has finite expectation is usually denoted by $EX < \infty$.

Example 2. Poisson distribution. Let X be Poisson distributed with parameter λ. Then

$$\begin{aligned} EX &= \sum_{j=1}^{\infty} j \frac{\lambda^j}{j!} e^{-\lambda} = \sum_{j=1}^{\infty} \frac{\lambda^j}{(j-1)!} e^{-\lambda} \\ &= \lambda e^{-\lambda} \sum_{j=0}^{\infty} \frac{\lambda^j}{j!} = \lambda e^{-\lambda} e^{\lambda} = \lambda. \end{aligned}$$

Example 3. Geometric distribution. Let X have a geometric distribution with parameter p. Find EX.

Now

$$\begin{aligned} EX &= \sum_{j=0}^{\infty} jp(1-p)^j \\ &= p(1-p) \sum_{j=0}^{\infty} j(1-p)^{j-1} \\ &= -p(1-p) \sum_{j=0}^{\infty} \frac{d}{dp} (1-p)^j. \end{aligned}$$

Since a power series can be differentiated term by term, it follows that

$$EX = -p(1-p) \frac{d}{dp} \sum_{j=0}^{\infty} (1-p)^j.$$

Using the formula for the sum of a geometric progression, we see that

$$EX = -p(1-p) \frac{d}{dp}\left(\frac{1}{p}\right) = -p(1-p)\left(\frac{-1}{p^2}\right).$$

Consequently

$$EX = \frac{1 - p}{p}.$$

We will next consider an example of a density that does not have finite mean.

Example 4. Let f be the function defined on R by

$$f(x) = \begin{cases} \dfrac{1}{x(x+1)}, & x = 1, 2, \ldots, \\ 0, & \text{elsewhere.} \end{cases}$$

The function f obviously satisfies properties (i) and (ii) in the definition of density functions given in Chapter 3. To see that f satisfies property (iii) we note that

$$\frac{1}{x(x+1)} = \frac{1}{x} - \frac{1}{x+1}$$

and hence

$$\sum_{x=1}^{\infty} f(x) = \sum_{x=1}^{\infty} \left[\frac{1}{x} - \frac{1}{x+1}\right]$$
$$= (1 - 1/2) + (1/2 - 1/3) + \cdots = 1.$$

Thus (iii) holds and f is a density. Now f does not have finite mean because

$$\sum_{x=1}^{\infty} |x| f(x) = \sum_{x=1}^{\infty} \frac{1}{x+1}$$

and it is well known that the harmonic series $\sum_{x=1}^{\infty} x^{-1}$ does not converge.

4.2. Properties of expectation

Often we want to compute the expectation of a random variable such as $Z = X_1 + X_2$ or $Z = X^2$ that is itself a function $\varphi(\mathbf{X})$ of the random vector $\mathbf{X}$. Of course, if we know the density f_Z of Z, this can be done by using (2). Quite frequently, however, the density of Z may not be known, or the computation of EZ from a known density of Z may be quite difficult. Our next result will give us a way of deciding if Z has finite expectation and, if it does, of computing EZ directly in terms of the density $f_{\mathbf{X}}$ and the function φ.

Before stating this result we introduce a notational convention. Let $\mathbf{X}$ be a discrete r-dimensional random vector having possible values $\mathbf{x}_1, \mathbf{x}_2, \ldots$ and density f, and let φ be any real-valued function defined on R^r. Then $\sum_{\mathbf{x}} \varphi(\mathbf{x}) f(\mathbf{x})$ is defined as

$$\sum_{\mathbf{x}} \varphi(\mathbf{x}) f(\mathbf{x}) = \sum_{j} \varphi(\mathbf{x}_j) f(\mathbf{x}_j). \tag{3}$$

Theorem 1 *Let* $\mathbf{X}$ *be a discrete random vector having density* f, *and let* φ *be a real-valued function defined on* R^r. *Then the random variable* $Z = \varphi(\mathbf{X})$ *has finite expectation if and only if*

$$\sum_{\mathbf{x}} |\varphi(\mathbf{x})| f(\mathbf{x}) < \infty \tag{4}$$

and, when (4) *holds,*

$$EZ = \sum_{\mathbf{x}} \varphi(\mathbf{x}) f(\mathbf{x}). \tag{5}$$

Proof. Let $z_1, z_2, \ldots$ denote the distinct possible values of Z and let $\mathbf{x}_1, \mathbf{x}_2, \ldots$ denote the distinct possible values of $\mathbf{X}$. For any z_j there is at least one $\mathbf{x}_i$ such that $z_j = \varphi(\mathbf{x}_i)$, but there may be more than one such $\mathbf{x}_i$. Let A_j denote the collection of such $\mathbf{x}_i$'s, that is,

$$A_j = \{\mathbf{x}_i \mid \varphi(\mathbf{x}_i) = z_j\}.$$

Then $\{\mathbf{X} \in A_j\}$ and $\{Z = z_j\}$ denote exactly the same events. Thus

$$P(Z = z_j) = P(\mathbf{X} \in A_j) = \sum_{\mathbf{x} \in A_j} f_{\mathbf{X}}(\mathbf{x}).$$

Consequently,

$$\begin{aligned} \sum_j |z_j| f_Z(z_j) &= \sum_j |z_j| P(Z = z_j) \\ &= \sum_j |z_j| \sum_{\mathbf{x} \in A_j} f_{\mathbf{X}}(\mathbf{x}) \\ &= \sum_j \sum_{\mathbf{x} \in A_j} |z_j| f_{\mathbf{X}}(\mathbf{x}). \end{aligned}$$

Since $\varphi(\mathbf{x}) = z_j$ for $\mathbf{x}$ in A_j, it follows that

$$\sum_j |z_j| f_Z(z_j) = \sum_j \sum_{\mathbf{x} \in A_j} |\varphi(\mathbf{x})| f_{\mathbf{X}}(\mathbf{x}).$$

By their definition, the sets A_j are disjoint for distinct values of j, and their union is the set of all possible values of $\mathbf{X}$. Therefore

$$\sum_j |z_j| f_Z(z_j) = \sum_{\mathbf{x}} |\varphi(\mathbf{x})| f_{\mathbf{X}}(\mathbf{x}).$$

This shows that Z has finite expectation if and only if (4) holds.

If Z does have finite expectation, then by repeating the above argument with the absolute signs eliminated, we conclude that (5) holds. ∎

Let X be a random variable having density f and let $\varphi(x) = |x|$. Then by Theorem 1, $|X|$ has finite expectation if and only if $\sum_x |x| f(x) < \infty$. But, according to our definition of expectation, X has finite expectation if and only if the same series converges. We see therefore that X has finite expectation if and only if $E|X| < \infty$.

We shall now use Theorem 1 to establish the following important properties of expectation.

Theorem 2 *Let X and Y be two random variables having finite expectation.*

(i) *If c is a constant and $P(X = c) = 1$, then $EX = c$.*

(ii) *If c is a constant, then cX has finite expectation and $E(cX) = cEX$.*

(iii) *$X + Y$ has finite expectation and*

$$E(X + Y) = EX + EY.$$

(iv) *Suppose $P(X \geq Y) = 1$. Then $EX \geq EY$; moreover, $EX = EY$ if and only if $P(X = Y) = 1$.*

(v) *$|EX| \leq E|X|$.*

Proof. The proof of (i) is quite simple. If $P(X = c) = 1$, then X has density $f_X(x) = 0$ for $x \neq c$ and $f_X(c) = 1$. Thus by (2)

$$EX = \sum_x x f_X(x) = c f_X(c) = c.$$

To prove (ii) let $\varphi(x) = cx$ and observe that

$$\sum_x |cx| f_X(x) = |c| \sum_x |x| f_X(x) < \infty,$$

so cX has finite expectation. Thus by (5)

$$E(cX) = \sum_x (cx) f_X(x) = c \sum_x x f_X(x) = cEX.$$

To establish (iii) set $\varphi(x, y) = x + y$ and let f denote the joint density of X and Y. Then

$$\begin{aligned}\sum_{x,y} |x + y| f(x, y) &\leq \sum_{x,y} |x| f(x, y) + \sum_{x,y} |y| f(x, y) \\ &= \sum_x |x| \sum_y f(x, y) + \sum_y |y| \sum_x f(x, y) \\ &= \sum_x |x| f_X(x) + \sum_y |y| f_Y(y) < \infty\end{aligned}$$

and hence $X + Y$ has finite expectation. Applying (5) we see that

$$\begin{aligned}E(X + Y) &= \sum_{x,y} (x + y) f(x, y) \\ &= \sum_{x,y} x f(x, y) + \sum_{x,y} y f(x, y) \\ &= EX + EY.\end{aligned}$$

To prove (iv) observe that $Z = X - Y = X + (-Y)$, and by (ii) and (iii) we see that

$$EX - EY = E(X - Y) = EZ = \sum_z z f_Z(z).$$

Since $P(Z \geq 0) = P(X \geq Y) = 1$, the values z_i that $Z = X - Y$ assumes must all be nonnegative. Thus $\sum_z z f_Z(z) \geq 0$ and hence $EX - EY \geq 0$. This yields the first part of (iv). If $EX = EY$, then $EZ = 0$. But then

$$0 = EZ = \sum_i z_i f_Z(z_i).$$

Now the sum of nonnegative terms can equal zero only if all the individual terms equal zero. Since $f_Z(z_i) > 0$ it must be that $z_i = 0$. Thus the only possible value of Z is 0 and consequently $P(Z = 0) = 1$.

Finally, (v) follows from (iv) and (ii) because $-|X| \leq X \leq |X|$ and hence $-E|X| \leq EX \leq E|X|$. This completes the proof of the theorem. ▮

It easily follows from (ii) and (iii) that if $X_1, \ldots, X_n$ are any n random variables having finite expectation, and $c_1, \ldots, c_n$ are any n constants, then

(6) $$E(c_1X_1 + \cdots + c_nX_n) = c_1EX_1 + \cdots + c_nEX_n.$$

It is useful to know that a bounded random variable always has finite expectation. More precisely,

Theorem 3 *Let X be a random variable such that for some constant M, $P(|X| \leq M) = 1$. Then X has finite expectation and $|EX| \leq M$.*

Proof. Let $x_1, x_2, \ldots$ be the possible values of X. Then $|x_i| \leq M$ for all i. Indeed, if $|x_i| > M$ for some possible value x_i, then

$$P(|X| > M) \geq P(|X| = |x_i|) > 0,$$

which contradicts the fact that $P(|X| \leq M) = 1$. Consequently

$$\sum_i |x_i| f(x_i) \leq M \sum_i f(x_i) \leq M,$$

so X has finite expectation. Moreover by (v) of Theorem 2,

$$|EX| \leq E|X| = \sum_i |x_i| f(x_i) \leq M.$$

This completes the proof. ▮

It follows easily from Theorem 3 and (iii) of Theorem 2 that if X and Y are two random variables such that Y has finite expectation and for some constant M, $P(|X - Y| \leq M) = 1$, then X also has finite expectation and $|EX - EY| \leq M$. We leave the proof of this fact for the reader.

Since the expectation of the sum of two random variables is the sum of their expectations, one might suppose that the expectation of a product is the product of the expectations. That this is not true in general can be seen by considering the random variable X taking values 1 and -1 each with probability 1/2 and setting $Y = X$. Then $EX = EY = 0$ but $EXY = EX^2 = 1$.

There is an important case when this product rule is valid, namely, when X and Y are independent random variables. We state this more formally as follows.

Theorem 4 *Let X and Y be two independent random variables having finite expectations. Then XY has finite expectation and*

$$E(XY) = (EX)(EY). \tag{7}$$

Proof. Observe that since X and Y are independent, the joint density of X and Y is $f_X(x)f_Y(y)$. Thus

$$\begin{aligned}\sum_{x,y} |xy| f(x, y) &= \sum_{x,y} |x|\,|y| f_X(x) f_Y(y) \\ &= \left(\sum_x |x| f_X(x)\right)\left(\sum_y |y| f_Y(y)\right) < \infty,\end{aligned}$$

so XY has finite expectation. Using Theorem 1, we conclude that

$$\begin{aligned}E(XY) &= \sum_{x,y} (xy) f_X(x) f_Y(y) \\ &= \left[\sum_x x f_X(x)\right]\left[\sum_y y f_Y(y)\right] = (EX)(EY).\end{aligned}$$ ∎

The converse of this property does not hold; two random variables X and Y may be such that $E(XY) = (EX)(EY)$ even though X and Y are not independent.

Example 5. Let (X, Y) assume the values $(1, 0)$, $(0, 1)$, $(-1, 0)$, and $(0, -1)$ with equal probabilities. Then $EX = EY = 0$. Since $XY = 0$, it follows that $E(XY) = 0$ and hence $E(XY) = (EX)(EY)$. To see that X and Y are not independent observe, for example, that $P(X = 0) = P(Y = 0) = 1/2$ whereas $P(X = 0, Y = 0) = 0$. Thus

$$P(X = 0, Y = 0) \neq P(X = 0)P(Y = 0).$$

It is often easier to compute expectations by using the properties given in Theorem 2 than by using the definition directly. We now illustrate this technique with several examples.

Example 6. Binomial distribution. We already know from Example 1 that the mean of the binomial distribution with parameters n and p is np. We can also derive this fact in a very simple manner by using the property that the expectation of a sum is the sum of the expectations ((iii) of Theorem 2). To this end, let $X_1, \ldots, X_n$ be n independent Bernoulli random variables having parameter p and set $S_n = X_1 + \cdots + X_n$. Then S_n has the binomial distribution with parameters n and p. By the first part of Example 1, $EX_i = p$, $1 \le i \le n$, and hence

$$ES_n = E(X_1 + \cdots + X_n) = \sum_{i=1}^{n} EX_i = np.$$

Example 7. Hypergeometric distribution. Suppose we have a population of r objects, r_1 of which are of type one and $r - r_1$ of type two. A sample of size n is drawn without replacement from this population. Let S_n denote the number of objects of type one that are obtained. Compute ES_n.

We know that S_n has the hypergeometric distribution, so we could compute ES_n by using (2). It is far simpler, however, to proceed by introducing indicator random variables $X_1, \ldots, X_n$ as follows. The random variable $X_i = 1$ if and only if the ith element in the sample is of type one. Then

$$EX_i = P(X_i = 1) = \frac{r_1}{r}.$$

But $S_n = X_1 + \cdots + X_n$, so using (iii) of Theorem 2 we see that

$$ES_n = \sum_{i=1}^{n} EX_i = n\frac{r_1}{r}.$$

Note that the random variables X_i, $1 \le i \le n$, are *not* independent.

Example 8. Suppose we have a population of r distinct objects labeled $1, 2, \ldots, r$. Objects are drawn with replacement until exactly $k \le r$ distinct objects have been obtained. Let S_k denote the size of the sample required. Compute ES_k.

It is clear that $S_1 = 1$ and hence $ES_1 = 1$. Assume $k \ge 2$ and let $X_i = S_{i+1} - S_i$, $i = 1, 2, \ldots, k - 1$. Then clearly $S_k = 1 + X_1 + \cdots + X_{k-1}$. Now X_i is the number of objects that must be drawn after the ith new object enters the sample and until the $(i + 1)$st new object enters the sample. A moment's thought shows that the event $\{X_i = n\}$ occurs if and only if the first $n - 1$ objects drawn after the ith new object enters the sample duplicate one of the previous i objects, and the nth object drawn after the ith new object enters is different from the previous i objects. Thus, as the trials are independent,

$$P(X_i = n) = \left(\frac{i}{r}\right)^{n-1}\left(1 - \frac{i}{r}\right), \qquad n = 1, 2, \ldots.$$

This shows that the random variable $X_i - 1$ is geometric with parameter $p_i = 1 - (i/r)$. Hence by Example 3, $E(X_i - 1) = p_i^{-1}(1 - p_i)$, and

$$EX_i = p_i^{-1}(1 - p_i) + 1 = p_i^{-1} = (1 - i/r)^{-1} = r(r - i)^{-1}.$$

Consequently,

$$ES_k = 1 + \sum_{i=1}^{k-1} \left(\frac{r}{r-i}\right) \tag{8}$$

$$= \sum_{i=0}^{k-1} \left(\frac{r}{r-i}\right)$$

$$= r\left(\frac{1}{r} + \frac{1}{r-1} + \cdots + \frac{1}{r-k+1}\right).$$

We point out for later use that it is clear from the construction of the X_i that they are mutually independent random variables.

In the previous chapter we have seen that nonnegative integer-valued random variables X play a prominent role. For these random variables the following theorem can frequently be applied both to decide if X has finite expectation and to compute the expectation of X.

Theorem 5 *Let X be a nonnegative integer-valued random variable. Then X has finite expectation if and only if the series $\sum_{x=1}^{\infty} P(X \geq x)$ converges. If the series does converge, then*

$$EX = \sum_{x=1}^{\infty} P(X \geq x). \tag{9}$$

Proof. We will show that

$$\sum_{x=1}^{\infty} xP(X = x) = \sum_{x=1}^{\infty} P(X \geq x), \tag{10}$$

from which the theorem follows immediately. To this end we first write the left side of (10) as

$$\sum_{x=1}^{\infty} P(X = x) \sum_{y=1}^{x} 1.$$

It is permissible to interchange the order of summation and rewrite this expression as

$$\sum_{y=1}^{\infty} \sum_{x=y}^{\infty} P(X = x) = \sum_{y=1}^{\infty} P(X \geq y).$$

Replacing the dummy variable y by the dummy variable x in the right side of this equality, we obtain the right side of (10). This shows that (10) holds, as desired. ∎

For an elementary application of this theorem, suppose that X is a geometrically distributed random variable having parameter p. Then $P(X \geq x) = (1 - p)^x$ and thus by the theorem

$$EX = \sum_{x=1}^{\infty} (1 - p)^x = (1 - p) + (1 - p)^2 = \cdots = p^{-1}(1 - p).$$

This agrees with the result found in Example 3.

4.3. Moments

Let X be a discrete random variable, and let $r \geq 0$ be an integer. We say that X has a moment of order r if X^r has finite expectation. In that case we define the rth moment of X as EX^r. If X has a moment of order r then the rth moment of $X - \mu$, where μ is the mean of X, is called the rth *central moment* (or the rth moment about the mean) of X. By Theorem 1 we can compute the rth moment and the rth central moment of X directly from the density f by the formulas

$$EX^r = \sum_x x^r f(x) \tag{11}$$

and

$$E(X - \mu)^r = \sum_x (x - \mu)^r f(x). \tag{12}$$

In view of (11) and (12), the rth moment and rth central moment are determined by the density f, and it therefore makes perfectly good sense to speak of them as the rth moment and rth central moment of this density.

Suppose X has a moment of order r; then X has a moment of order k for all $k \leq r$. To see this, observe that if $|x| \leq 1$, then

$$|x^k| = |x|^k \leq 1$$

while for $|x| > 1$,

$$|x|^k \leq |x|^r.$$

Thus in either case it is always true that

$$|x|^k \leq |x|^r + 1.$$

Thus, by the comparison theorem for the convergence of series, we see that

$$\sum_x |x|^k f(x) \leq \sum_x [|x|^r + 1] f(x) = E(|X|^r) + 1 < \infty,$$

so X^k has finite expectation.

On the other hand, as was shown in Example 4, a random variable X may not have even a first moment. A simple modification of this example shows that a random variable may have a moment of order r but possess no higher order moment. (See Exercise 9.)

The first moment ($r = 1$) is just the mean of X. In general, the more moments of X that we know the more information we have gained about the distribution of X; however, in applications it is the first two moments that are usually of major interest.

By property (iii) of Theorem 2 we know that if X and Y both have a finite first moment, then so does $X + Y$. We will now show that this desirable property holds true for moments of order r as well.

Theorem 6 *If the random variables X and Y have moments of order r, then X + Y also has a moment of order r.*

Proof. This theorem rests on the following simple inequality. For any nonnegative integer $j \leq r$,

$$|x|^j |y|^{r-j} \leq |x|^r + |y|^r, \qquad x, y \in R. \tag{13}$$

To see this, observe that if $|x| \leq |y|$, then $|x|^j |y|^{r-j} \leq |y|^j |y|^{r-j} = |y|^r \leq |x|^r + |y|^r$; while if $|x| \geq |y|$, then $|x|^j |y|^{r-j} \leq |x|^r \leq |x|^r + |y|^r$. Thus (13) holds. Using (13) and the binomial theorem we now see that

$$\begin{aligned} |x + y|^r &\leq (|x| + |y|)^r \\ &= \sum_{j=0}^{r} \binom{r}{j} |x|^j |y|^{r-j} \\ &\leq \sum_{j=0}^{r} \binom{r}{j} (|x|^r + |y|^r). \end{aligned}$$

But

$$\sum_{j=0}^{r} \binom{r}{j} = 2^r$$

because

$$2^r = (1 + 1)^r = \sum_{j=0}^{r} \binom{r}{j} 1^j 1^{r-j} = \sum_{j=0}^{r} \binom{r}{j}.$$

Consequently

$$|x + y|^r \leq 2^r(|x|^r + |y|^r).$$

Let f be the joint density of X and Y. Then

$$\begin{aligned} \sum_{x,y} |x + y|^r f(x, y) &\leq 2^r \sum_{x,y} (|x|^r + |y|^r) f(x, y) \\ &= 2^r E(|X|^r + |Y|^r) \\ &= 2^r (E|X|^r + E|Y|^r) < \infty. \end{aligned}$$

Hence by Theorem 1, $(X + Y)^r$ has finite expectation. ∎

It follows easily by induction that if $X_1, X_2, \ldots, X_n$ all have a moment of order r, then so does $X_1 + \cdots + X_n$.

Let X be a random variable having a finite second moment. Then the *variance of X*, denoted by $\operatorname{Var} X$ or $V(X)$, is defined by

$$\operatorname{Var} X = E[(X - EX)^2].$$

By expanding the right-hand side we see that

$$\begin{aligned} \operatorname{Var} X &= E[X^2 - (2X)(EX) + (EX)^2] \\ &= EX^2 - 2(EX)^2 + (EX)^2. \end{aligned}$$

In other words

$$\text{(14)} \qquad \operatorname{Var} X = EX^2 - (EX)^2.$$

One often denotes EX by μ and $\operatorname{Var} X$ by σ^2. The nonnegative number $\sigma = \sqrt{\operatorname{Var} X}$ is called the *standard deviation* of X or of f_X.

According to our previous discussion, the mean μ is the average value of the random variable X. One use of the variance is as a measure of the spread of the distribution of X about the mean. The more X tends to deviate from its mean value μ, the larger $(X - \mu)^2$ tends to be, and hence the larger the variance becomes.

On the other hand, $\operatorname{Var} X = 0$ if and only if X is a constant. To see this, observe that if $P(X = c) = 1$ for some constant c, then $EX = c$ and $\operatorname{Var} X = 0$. Conversely, if $\operatorname{Var} X = 0$, then $E[(X - EX)^2] = 0$ and hence $P((X - EX)^2 = 0) = 1$. Consequently $P(X = EX) = 1$.

An alternative use of the mean and variance is provided by the following problem, which is of interest in statistics. Let X be a random variable having a finite second moment, and suppose we want to choose the value of a that minimizes $E(X - a)^2$. Such a value would provide the best fit to X by a constant if the error were measured by the mean square deviation.

One way of solving this problem is to use calculus. Note that

$$E(X - a)^2 = EX^2 - 2aEX + a^2.$$

If we differentiate with respect to a and set the derivative equal to zero, we see that $a = EX$. Since the second derivative is positive (in fact, it equals 2), the point corresponds to a minimum, and the minimum value is $\operatorname{Var} X$.

There is a second way of solving this problem that is also important to understand. Note that

$$\begin{aligned}(X - a)^2 &= [(X - \mu) + (\mu - a)]^2 \\ &= (X - \mu)^2 + 2(X - \mu)(\mu - a) + (\mu - a)^2.\end{aligned}$$

Since $E(X - \mu) = 0$, it follows that the cross-product term has zero expectation and hence

$$\text{(15)} \qquad \begin{aligned}E(X - a)^2 &= E(X - \mu)^2 + (\mu - a)^2 \\ &= \operatorname{Var} X + (\mu - a)^2.\end{aligned}$$

It is now clear from (15) that $E(X - a)^2$ is at a minimum when $\mu = a$, and this minimum value is $\operatorname{Var} X$.

We can often find the moments of a nonnegative integer-valued random variable X most simply by differentiating its probability generating function Φ_X. Suppose for simplicity that

$$\sum_{x=0}^{\infty} f_X(x)t_0^x < \infty$$

for some $t_0 > 1$. Then we can regard Φ_X as being defined on $-t_0 < t < t_0$ by

$$\Phi_X(t) = \sum_{x=0}^{\infty} f_X(x)t^x, \qquad -t_0 < t < t_0.$$

We can differentiate $\Phi_X(t)$ any number of times by differentiating the corresponding power series term by term. In particular

$$\Phi_X'(t) = \sum_{x=1}^{\infty} x f_X(x)t^{x-1}, \qquad -t_0 < t < t_0,$$

and

$$\Phi_X''(t) = \sum_{x=2}^{\infty} x(x-1) f_X(x)t^{x-2}, \qquad -t_0 < t < t_0.$$

By our assumptions on t_0, we can let $t = 1$ in these formulas, obtaining

$$\Phi_X'(1) = \sum_{x=1}^{\infty} x f_X(x) = EX$$

and

$$\Phi_X''(1) = \sum_{x=2}^{\infty} x(x-1) f_X(x) = EX(X-1).$$

Thus the mean and variance of X can be obtained from Φ_X by means of the formulas

$$EX = \Phi_X'(1)$$

and

$$\operatorname{Var} X = EX^2 - (EX)^2 = \Phi_X''(1) + \Phi_X'(1) - (\Phi_X'(1))^2.$$

Similar formulas, in terms of the higher derivatives of $\Phi_X(t)$ at $t = 1$, can be developed for the other moments of X.

We now illustrate the use of these formulas with the following examples.

Example 9. Negative binomial distribution. Let X be a random variable having a negative binomial distribution with parameters α and p. Find the mean and variance of X.

From Example 17 of Chapter 3, we know that the generating function of X is given by $\Phi_X(t) = p^\alpha[1 - t(1-p)]^{-\alpha}$. Consequently,

$$\Phi_X'(t) = \alpha p^\alpha[1 - t(1-p)]^{-(\alpha+1)}(1-p)$$

and

$$\Phi_X''(t) = (\alpha+1)\alpha p^\alpha[1 - t(1-p)]^{-(\alpha+2)}(1-p)^2.$$

Thus

$$\Phi_X'(1) = \alpha\left(\frac{1-p}{p}\right)$$

and

$$\Phi_X''(1) = (\alpha+1)\alpha\left(\frac{1-p}{p}\right)^2.$$

Hence, $EX = \alpha p^{-1}(1 - p)$ and

$$\begin{aligned}\text{Var } X &= (\alpha + 1)\alpha\left(\frac{1 - p}{p}\right)^2 + \alpha\left(\frac{1 - p}{p}\right) - \alpha^2\left(\frac{1 - p}{p}\right)^2 \\ &= \alpha\,\frac{1 - p}{p^2}.\end{aligned}$$

In particular, if X has a geometric distribution with parameter p, then $EX = p^{-1}(1 - p)$ (as we have already seen) and $\text{Var } X = p^{-2}(1 - p)$.

Example 10. Poisson distribution. Let X have a Poisson distribution with parameter λ. Find the mean and variance of X.

In Example 18 of Chapter 3 we found that $\Phi_X(t) = e^{\lambda(t-1)}$. Thus

$$\Phi_X'(t) = \lambda e^{\lambda(t-1)}$$

and

$$\Phi_X''(t) = \lambda^2 e^{\lambda(t-1)}.$$

Consequently $\Phi_X'(1) = \lambda$ and $\Phi_X''(1) = \lambda^2$. It follows immediately that

$$EX = \lambda,$$

which agrees with the answer found in Example 2, and

$$\text{Var } X = \lambda^2 + \lambda - \lambda^2 = \lambda.$$

This shows that if X has a Poisson distribution with parameter λ, then the mean and variance of X both equal λ.

4.4. Variance of a sum

Let X and Y be two random variables each having finite second moment. Then $X + Y$ has finite second moment and hence finite variance. Now

$$\begin{aligned}\text{Var }(X + Y) &= E[(X + Y) - E(X + Y)]^2 \\ &= E[(X - EX) + (Y - EY)]^2 \\ &= E(X - EX)^2 + E(Y - EY)^2 \\ &\quad + 2E[(X - EX)(Y - EY)] \\ &= \text{Var } X + \text{Var } Y + 2E[(X - EX)(Y - EY)].\end{aligned}$$

Thus, unlike the mean, the variance of a sum of two random variables is, in general, not the sum of the variances. The quantity

$$E[(X - EX)(Y - EY)]$$

is called the *covariance* of X and Y and written $\text{Cov }(X, Y)$. Thus we have the important formula

$$\text{Var }(X + Y) = \text{Var } X + \text{Var } Y + 2\,\text{Cov }(X, Y). \tag{16}$$

Now

$$(X - EX)(Y - EY) = XY - (Y)(EX) - X(EY) + (EX)(EY),$$

and hence taking expectations we see that

$$(17) \quad \operatorname{Cov}(X, Y) = E[(X - EX)(Y - EY)] = E(XY) - (EX)(EY).$$

From this form, it is clear that $\operatorname{Cov}(X, Y) = 0$ whenever X and Y are independent. (Example 5 shows that the converse is false.) We see from (16) that if X and Y are independent random variables having finite second moments, then $\operatorname{Var}(X + Y) = \operatorname{Var} X + \operatorname{Var} Y$.

In particular if $P(Y = c) = 1$ for some constant c, then X and Y are independent and the variance of Y equals zero; consequently

$$(18) \qquad \operatorname{Var}(X + c) = \operatorname{Var} X + \operatorname{Var}(c) = \operatorname{Var} X.$$

More generally, if $X_1, X_2, \ldots, X_n$ are n random variables each having a finite second moment, then

$$(19) \quad \operatorname{Var}\left(\sum_{i=1}^{n} X_i\right) = \sum_{i=1}^{n} \operatorname{Var} X_i + 2 \sum_{i=1}^{n-1} \sum_{j=i+1}^{n} \operatorname{Cov}(X_i, X_j),$$

and, in particular, if $X_1, \ldots, X_n$ are mutually independent, then

$$(20) \qquad \operatorname{Var}\left(\sum_{i=1}^{n} X_i\right) = \sum_{i=1}^{n} \operatorname{Var} X_i.$$

These formulas can be derived by a direct computation similar to (but more complicated than) that used for the case $n = 2$, or they can be established from the case $n = 2$ by induction on n.

In particular, if $X_1, \ldots, X_n$ are independent random variables having a common variance σ^2 (for example, if they each had the same density), then

$$(21) \qquad \operatorname{Var}(X_1 + \cdots + X_n) = n \operatorname{Var} X_1 = n\sigma^2.$$

Another elementary but quite useful fact is that $\operatorname{Var}(aX) = a^2 \operatorname{Var} X$. We leave the verification of this fact to the reader.

Example 11. Binomial distribution. Let $X_1, \ldots, X_n$ be n independent Bernoulli random variables each having the same probability p of assuming the value 1. Then (see Example 6) the sum $S_n = X_1 + \cdots + X_n$ is binomially distributed with parameters n and p. We have previously shown that $ES_n = np$. Using (21) we find at once that

$$\operatorname{Var} S_n = n \operatorname{Var} X_1.$$

Now $X_1^2 = X_1$ because X_1 is either 0 or 1. Thus $EX_1^2 = EX_1 = p$ and hence

$$\operatorname{Var} X_1 = EX_1^2 - (EX_1)^2 = p - p^2 = p(1 - p).$$

Consequently $\operatorname{Var} S_n = np(1 - p)$.

In summary then, the mean of a binomially distributed variable is np and its variance is $np(1 - p)$.

Example 12. Hypergeometric distribution. Consider the situation in Example 7. We now want to compute $\operatorname{Var} S_n$, so as to obtain the variance of a hypergeometric distribution. To this end we will use Equation (19).

For the dependent indicators, $X_1, \ldots, X_n$, we previously found that

$$P(X_i = 1) = EX_i = \frac{r_1}{r}.$$

Since $X_i^2 = X_i$ we see that

$$\operatorname{Var} X_i = EX_i^2 - (EX_i)^2 = \left(\frac{r_1}{r}\right) - \left(\frac{r_1}{r}\right)^2$$

$$= \left(\frac{r_1}{r}\right)\left(1 - \frac{r_1}{r}\right).$$

Next we must compute the covariances. Assume that $1 \le i < j \le n$. Now $X_iX_j = 0$ unless both X_i and X_j are 1, so

$$EX_iX_j = P(X_i = 1, X_j = 1) = \left(\frac{r_1}{r}\right)\left(\frac{r_1 - 1}{r - 1}\right).$$

Thus

$$\operatorname{Cov}(X_i, X_j) = E(X_iX_j) - (EX_i)(EX_j)$$

$$= \frac{r_1(r_1 - 1)}{r(r - 1)} - \left(\frac{r_1}{r}\right)^2$$

$$= \left(\frac{r_1}{r}\right)\left(\frac{r_1 - 1}{r - 1} - \frac{r_1}{r}\right)$$

$$= \left(\frac{r_1}{r}\right)\frac{r_1 - r}{r(r - 1)},$$

and hence

$$\sum_{i=1}^{n-1}\sum_{j=i+1}^{n} \operatorname{Cov}(X_i, X_j) = \frac{n(n - 1)}{2}\left(\frac{r_1}{r}\right)\frac{r_1 - r}{r(r - 1)}.$$

It now follows from (19) that

$$\operatorname{Var} S_n = n\frac{r_1(r - r_1)}{r^2} - n(n - 1)\frac{r_1(r - r_1)}{r^2(r - 1)}$$

$$= n\left(\frac{r_1}{r}\right)\left(1 - \frac{r_1}{r}\right)\left(1 - \frac{n - 1}{r - 1}\right).$$

It is interesting to compare the mean and variance for the hypergeometric distribution with those of the binomial distribution having the same success probability $p = (r_1/r)$. Suppose we have a population of r objects of which r_1 are of type one and $r - r_1$ are of type two. A random sample of size n is drawn from the population. Let Y denote the number of objects of type one in the sample.

If the sampling is done *with replacement* then Y is binomially distributed with parameters n and $p = (r_1/r)$, and hence

$$EY = n\left(\frac{r_1}{r}\right) \quad \text{and} \quad \operatorname{Var} Y = n\left(\frac{r_1}{r}\right)\left(1 - \frac{r_1}{r}\right).$$

On the other hand, if the sampling is done *without replacement*, then Y has a hypergeometric distribution,

$$EY = n\left(\frac{r_1}{r}\right) \quad \text{and} \quad \operatorname{Var} Y = n\left(\frac{r_1}{r}\right)\left(1 - \frac{r_1}{r}\right)\left(1 - \frac{n-1}{r-1}\right).$$

The mean is the same in the two cases, but in sampling without replacement the variance is smaller. Intuitively, the closer n is to r the more deterministic Y becomes when we sample without replacement. Indeed, if $n = r$ then the variance is zero and $P(Y = r_1) = 1$. But if r is large compared to n, so that (n/r) is close to zero, the ratio of the variances obtained in sampling with and without replacements is close to one. This is as it should be, since for fixed n and large r there is little difference between sampling with replacement and sampling without replacement.

4.5. Correlation coefficient

Let X and Y be two random variables having finite nonzero variances. One measure of the degree of dependence between the two random variables is the correlation coefficient $\rho(X, Y)$ defined by

$$\rho = \rho(X, Y) = \frac{\operatorname{Cov}(X, Y)}{\sqrt{(\operatorname{Var} X)(\operatorname{Var} Y)}}. \tag{22}$$

These random variables are said to be uncorrelated if $\rho = 0$. Since $\operatorname{Cov}(X, Y) = 0$ if X and Y are independent, we see at once that independent random variables are uncorrelated. It is also possible for dependent random variables to be uncorrelated, as can be seen from Example 5.

It is important for applications in statistics to know that the correlation coefficient ρ is always between -1 and 1, and that $|\rho| = 1$ if and only if $P(X = aY) = 1$ for some constant a. These facts are easy consequences of the following basic inequality called the *Schwarz inequality*.

Theorem 7 The Schwarz Inequality. *Let X and Y have finite second moments. Then*

$$[E(XY)]^2 \le (EX^2)(EY^2). \tag{23}$$

Furthermore, equality holds in (23) *if and only if either $P(Y = 0) = 1$ or $P(X = aY) = 1$ for some constant a.*

Proof. If $P(Y = 0) = 1$, then $P(XY = 0) = 1$, $EXY = 0$, and $EY^2 = 0$; thus in this case (23) holds with equality. Also, if $P(X = aY) = 1$, then a simple computation will show that both sides of (23) are equal to $(a^2EY^2)^2$.

We now show that (23) always holds. From the above discussion we can assume that $P(Y = 0) < 1$ and hence $EY^2 > 0$. The proof is based on a simple but clever trick. Observe that for any real number λ

$$0 \leq E(X - \lambda Y)^2 = \lambda^2 EY^2 - 2\lambda EXY + EX^2.$$

This is a quadratic function of λ. Since the coefficient EY^2 of λ^2 is positive, the minimum is achieved for some value of λ, say $\lambda = a$, that can be found by the usual calculus method of setting the derivative equal to zero and solving. The answer is $a = [E(XY)][EY^2]^{-1}$. Since the corresponding function value is

$$0 \leq E(X - aY)^2 = EX^2 - \frac{[E(XY)]^2}{EY^2} \tag{24}$$

it follows that (23) holds. If equality holds in the Schwarz inequality (23), then from (24) we see that $E(X - aY)^2 = 0$, so that

$$P[(X - aY) = 0] = 1.$$

This completes the proof. ∎

Applying the Schwarz inequality to the random variables $(X - EX)$ and $(Y - EY)$ we see that

$$(E[(X - EX)(Y - EY)])^2 \leq [E(X - EX)^2][E(Y - EY)^2];$$

that is,

$$[\text{Cov}\,(X, Y)]^2 \leq (\text{Var}\,X)(\text{Var}\,Y).$$

Thus by the definition of ρ

$$|\rho(X, Y)| \leq 1.$$

We also see from Theorem 7 that $|\rho| = 1$ if and only if $P(X = aY) = 1$ for some constant a.

The correlation coefficient is of limited use in probability theory. It arises mainly in statistics and further discussion of it will be postponed to Volume II.

4.6. Chebyshev's inequality

Let X be a nonnegative random variable having finite expectation, and let t be a positive real number. Define the random variable Y by setting $Y = 0$ if $X < t$ and $Y = t$ if $X \geq t$. Then Y is a discrete random variable having the two possible values 0 and t which it assumes with probabilities $P(Y = 0) = P(X < t)$ and $P(Y = t) = P(X \geq t)$ respectively. Thus

$$EY = tP(Y = t) + 0 \cdot P(Y = 0) = tP(Y = t) = tP(X \geq t).$$

Now clearly $X \geq Y$ and hence $EX \geq EY$. Thus

$$EX \geq EY = tP(X \geq t)$$

or

$$P(X \geq t) \leq \frac{EX}{t}. \tag{25}$$

Quite a variety of useful inequalities can be deduced from (25). The most important of these is the Chebyshev inequality.

Chebyshev's Inequality. Let X be a random variable with mean μ and finite variance σ^2. Then for any real number $t > 0$

$$P(|X - \mu| \geq t) \leq \frac{\sigma^2}{t^2}. \tag{26}$$

To prove (26), we apply (25) to the nonnegative random variable $(X - \mu)^2$ and the number t^2. We conclude that

$$P((X - \mu)^2 \geq t^2) \leq \frac{E(X - \mu)^2}{t^2} = \frac{\sigma^2}{t^2}.$$

Since $(X - \mu)^2 \geq t^2$ if and only if $|X - \mu| \geq t$ we see that (26) holds.

Chebyshev's inequality gives an upper bound in terms of Var X and t for the probability that X deviates from its mean by more than t units. Its virtue lies in its great generality. No assumption on the distribution of X is made other than that it has finite variance. This inequality is the starting point for several theoretical developments. For most distributions that arise in practice, there are far sharper bounds for $P(|X - \mu| \geq t)$ than that given by Chebyshev's inequality; however, examples show that in general the bound given by Chebyshev's inequality cannot be improved upon (see Exercise 26).

Let $X_1, \ldots, X_n$ be n independent random variables having the same distribution. These random variables may be thought of as n independent measurements of some quantity that is distributed according to their common distribution. In this sense we sometimes speak of the random variables $X_1, \ldots, X_n$ as constituting a random sample of size n from this distribution.

Suppose that the common distribution of these random variables has finite mean μ. Then for n sufficiently large we would expect that their arithmetic mean $S_n/n = (X_1 + \cdots + X_n)/n$ should be close to μ. If the X_i also have finite variance, then

$$\operatorname{Var}\left(\frac{S_n}{n}\right) = \frac{n\sigma^2}{n^2} = \frac{\sigma^2}{n}$$

and thus $\operatorname{Var}(S_n/n) \to 0$ as $n \to \infty$. As discussed in Section 4.3, this implies that as n gets large the distribution of S_n/n becomes more concentrated about its mean μ. More precisely, by applying Chebyshev's inequality to S_n/n we obtain the inequality

$$P\left(\left|\frac{S_n}{n} - \mu\right| \geq \delta\right) \leq \frac{\operatorname{Var}(S_n/n)}{\delta^2} = \frac{\sigma^2}{n\delta^2}. \tag{27}$$

In particular, it follows from (27) that for any $\delta > 0$

$$\lim_{n\to\infty} P\left(\left|\frac{S_n}{n} - \mu\right| \geq \delta\right) = 0. \tag{28}$$

We may interpret (28) in the following way. The number δ can be thought of as the desired accuracy in the approximation of μ by S_n/n. Equation (28) assures us that no matter how small δ may be chosen the probability that S_n/n approximates μ to within this accuracy, that is, $P(|(S_n/n) - \mu| < \delta)$, converges to 1 as the number of observations gets large. This fact is called the Weak Law of Large Numbers. We have proven the weak law only under the assumption that the common variance of the X_i is finite. Actually this is not necessary; all that is needed is that the X_i have a finite mean. We state this more general result in the following theorem. The proof will be given in Chapter 8.

Theorem 8 Weak Law of Large Numbers. *Let $X_1, X_2, \ldots, X_n$ be independent random variables having a common distribution with finite mean μ and set $S_n = X_1 + \cdots + X_n$. Then for any $\delta > 0$*

$$\lim_{n\to\infty} P\left(\left|\frac{S_n}{n} - \mu\right| \geq \delta\right) = 0.$$

Whenever the X_i have finite mean, the weak law holds. However, when they also have finite variance, then (27) holds. This is a more precise statement since it gives us an upper bound for $P\left(\left|\frac{S_n}{n} - \mu\right| \geq \delta\right)$ in terms of n. We now illustrate the use of (27) by applying it to binomially distributed random variables.

Let $X_1, X_2, \ldots, X_n$ be n independent Bernoulli random variables assuming the value 1 with common probability p. Then $\mu = p$ and $\sigma^2 = p(1 - p)$. Thus (27) shows that

$$P\left(\left|\frac{S_n}{n} - p\right| \geq \delta\right) \leq \frac{p(1-p)}{n\delta^2}. \tag{29}$$

Since $p(1 - p) \leq 1/4$ if $0 < p < 1$ (because by the usual calculus methods it can easily be shown that $p(1 - p)$ has its maximum value at $p = 1/2$), it follows that regardless of what p may be,

$$P\left(\left|\frac{S_n}{n} - p\right| \geq \delta\right) \leq \frac{1}{4n\delta^2}. \tag{30}$$

Equation (29) is useful when we know the value of p, while (30) gives us a bound on $P\left(\left|\frac{S_n}{n} - p\right| \geq \delta\right)$ that is valid for any value of p. If p is near 1/2, (29) and (30) do not differ by much, but if p is far from 1/2 the estimate given by (29) may be much better. (Actually even the bounds given by (29) are quite poor. We shall discuss another method in Chapter 7 that yields much better estimates.)

Suppose δ and $\varepsilon > 0$ are given. We may use (29) or (30) to find a lower bound on the number of trials needed to assure us that

$$P\left(\left|\frac{S_n}{n} - p\right| \geq \delta\right) \leq \varepsilon.$$

Indeed, from (29) we see that this will be the case if $p(1 - p)/n\delta^2 \leq \varepsilon$. Solving for n we find that $n \geq p(1 - p)/\varepsilon\delta^2$. If we use (30), then $n \geq (4\varepsilon\delta^2)^{-1}$ trials will do. We state again that these bounds on n given by Chebyshev's inequality are poor and that in fact a much smaller number of trials may be sufficient.

As an illustration of the difference between these two estimates for n, choose $\delta = .1$ and $\varepsilon = .01$. Then $\delta^2\varepsilon = 10^{-4}$ and from (30) we see that to guarantee that

$$P\left(\left|\frac{S_n}{n} - p\right| \geq .1\right) \leq .01$$

we would need $n = 10^4/4 = 2500$ observations. Suppose, however, we knew that $p = .1$. Then since $p(1 - p) = .09$ we see from (29) that $n \geq .09 \times 10^4 = 900$ observations will suffice. For $p = 1/2$, (29) gives the same estimate as (30), namely 2500.

To illustrate that the Chebyshev bounds are really poor in the case of the binomial distribution, suppose $n = 100$ and $p = 1/2$. From (29) we then obtain

$$P\left(\left|\frac{S_n}{n} - .5\right| \geq .1\right) \leq .25.$$

This should be compared with the exact value for this probability which is .038.

Exercises

1 Let N be a positive integer and let f be the function defined by

$$f(x) = \begin{cases} \dfrac{2x}{N(N+1)}, & x = 1, 2, \ldots, N, \\ 0, & \text{elsewhere.} \end{cases}$$

Show that f is a discrete density and find its mean. *Hint:*

$$\sum_{x=1}^{N} x = \frac{N(N+1)}{2} \quad \text{and} \quad \sum_{x=1}^{N} x^2 = \frac{N(N+1)(2N+1)}{6}.$$

2 Let X have a binomial density with parameters $n = 4$ and p. Find $E \sin(\pi X/2)$.

3 Let X be Poisson with parameter λ. Compute the mean of $(1 + X)^{-1}$.

4 If X has mean 1 and Y has mean 3, what is the mean of $2X + 5Y$?

5 Suppose X and Y are two random variables such that

$$P(|X - Y| \le M) = 1$$

for some constant M. Show that if Y has finite expectation, then X has finite expectation and $|EX - EY| \le M$.

6 Let X be a geometrically distributed random variable and let $M > 0$ be an integer. Set $Z = \min(X, M)$. Compute the mean of Z. *Hint:* Use Theorem 5.

7 Let X be a geometrically distributed random variable and let $M > 0$ be an integer. Set $Y = \max(X, M)$. Compute the mean of Y. *Hint:* Compute $P(Y < y)$ and then use Theorem 5.

8 Let X be uniformly distributed on $\{0, 1, \ldots, N\}$. Find the mean and variance of X by using the hint to Exercise 1.

9 Construct an example of a density that has a finite moment of order r but has no higher finite moment. *Hint:* Consider the series $\sum_{x=1}^{\infty} x^{-(r+2)}$ and make this into a density.

10 Suppose X and Y are two independent random variables such that $EX^4 = 2$, $EY^2 = 1$, $EX^2 = 1$, and $EY = 0$. Compute $\operatorname{Var}(X^2 Y)$.

11 Show that $\operatorname{Var}(aX) = a^2 \operatorname{Var} X$.

12 Let X be binomially distributed with parameters n and p. Use the probability generating function of X to compute its mean and variance.

13 Let X be a nonnegative integer-valued random variable.

(a) Show that

$$\Phi_X(t) = Et^X, \qquad -1 \le t \le 1,$$

$$\Phi_X'(t) = EXt^{X-1}, \qquad -1 < t < 1,$$

and

$$\Phi_X''(t) = EX(X-1)t^{X-2}, \qquad -1 < t < 1.$$

(b) Use Theorem 4 to rederive the result that if X and Y are independent nonnegative integer-valued random variables, then

$$\Phi_{X+Y}(t) = \Phi_X(t)\Phi_Y(t), \qquad -1 \le t \le 1.$$

14 Let X and Y be two independent random variables having finite second moments. Compute the mean and variance of $2X + 3Y$ in terms of those of X and Y.

15 Let $X_1, \ldots, X_n$ be independent random variables having a common density with mean μ and variance σ^2. Set $\bar{X} = (X_1 + \cdots + X_n)/n$.
(a) By writing $X_k - \bar{X} = (X_k - \mu) - (\bar{X} - \mu)$, show that

$$\sum_{k=1}^{n} (X_k - \bar{X})^2 = \sum_{k=1}^{n} (X_k - \mu)^2 - n(\bar{X} - \mu)^2.$$

(b) Conclude from (a) that

$$E\left(\sum_{k=1}^{n} (X_k - \bar{X})^2\right) = (n - 1)\sigma^2.$$

16 Suppose n balls are distributed at random into r boxes. Let $X_i = 1$ if box i is empty and let $X_i = 0$ otherwise.
(a) Compute EX_i.
(b) For $i \ne j$, compute $E(X_iX_j)$.
(c) Let S_r denote the number of empty boxes. Write $S_r = X_1 + \cdots + X_r$, and use the result of (a) to compute ES_r.
(d) Use the result of (a) and (b) to compute Var S_r.

17 Suppose we have two decks of n cards, each numbered $1, \ldots, n$. The two decks are shuffled and the cards are matched against each other. We say a match occurs at position i if the ith card from each deck has the same number. Let S_n denote the number of matches.
(a) Compute ES_n.
(b) Compute Var S_n.
Hint: Let $X_i = 1$ if there is a match at position i and let $X_i = 0$ otherwise. Then $S_n = X_1 + \cdots + X_n$. From the results in Chapter 2 we know that $P(X_i = 1) = 1/n$ and that if $i \ne j$,

$$P(X_i = 1, X_j = 1) = \frac{1}{n(n - 1)}.$$

18 Consider the random variable S_k introduced in Example 8. Compute Var S_k.

19 Establish the following properties of covariance:
(a) $\operatorname{Cov}(X, Y) = \operatorname{Cov}(Y, X)$;
(b) $\operatorname{Cov}\left(\sum_{i=1}^{m} a_iX_i, \sum_{j=1}^{n} b_jY_j\right) = \sum_{i=1}^{m}\sum_{j=1}^{n} a_ib_j \operatorname{Cov}(X_i, Y_j).$

20 Let X_1, X_2, and X_3 be independent random variables having finite positive variances σ_1^2, σ_2^2, and σ_3^2 respectively. Find the correlation between $X_1 - X_2$ and $X_2 + X_3$.

21 Suppose X and Y are two random variables such that $\rho(X, Y) = 1/2$, $\operatorname{Var} X = 1$, and $\operatorname{Var} Y = 2$. Compute $\operatorname{Var}(X - 2Y)$.

22 A box has 3 red balls and 2 black balls. A random sample of size 2 is drawn without replacement. Let U be the number of red balls selected and let V be the number of black balls selected. Compute $\rho(U, V)$.

23 Suppose a box has 3 balls labeled 1, 2, and 3. Two balls are selected without replacement from the box. Let X be the number on the first ball and let Y be the number on the second ball. Compute $\operatorname{Cov}(X, Y)$ and $\rho(X, Y)$.

24 Suppose an experiment having r possible outcomes $1, 2, \ldots, r$ that occur with probabilities $p_1, \ldots, p_r$ is repeated n times. Let X be the number of times the first outcome occurs, and let Y be the number of times the second outcome occurs. Show that

$$\rho(X, Y) = -\sqrt{\frac{p_1 p_2}{(1 - p_1)(1 - p_2)}}$$

by carrying out the following steps. Let $I_i = 1$ if the ith trial yields outcome 1, and let $I_i = 0$ otherwise. Similarly, let $J_i = 1$ if the ith trial yields outcome 2, and let $J_i = 0$ elsewhere. Then $X = I_1 + \cdots + I_n$ and $Y = J_1 + \cdots + J_n$. Now show the following:

(a) $E(I_i J_i) = 0$.

(b) If $i \neq j$, $E(I_i J_j) = p_1 p_2$.

(c) $E(XY) = E\left(\sum_{i=1}^{n} I_i J_i\right) + E\left(\sum_{i=1}^{n} \sum_{j \neq i} I_i J_j\right)$
$= n(n - 1) p_1 p_2$.

(d) $\operatorname{Cov}(X, Y) = -n p_1 p_2$.

(e) $\rho(X, Y) = -\sqrt{\dfrac{p_1 p_2}{(1 - p_1)(1 - p_2)}}$.

25 Suppose a population of r objects consists of r_1 objects of type 1, r_2 objects of type 2, and r_3 objects of type 3, where $r_1 + r_2 + r_3 = r$. A random sample of size $n \leq r$ is selected without replacement from this population. Let X denote the number of objects of type 1 in the sample and let Y denote the number of objects of type 2. Compute $\rho(X, Y)$ by doing the following. Let $I_i = 1$ or 0 according as the ith element in the sample is of type 1 or not and let $J_i = 1$ or 0 according as the ith element in the sample is of type 2 or not.

(a) Show that $EI_i = r_1/r$ and $EJ_i = r_2/r$.

(b) Show that for $i \neq j$,

$$EI_i J_j = \frac{r_1 r_2}{r(r - 1)}$$

and that $E(I_i J_i) = 0$.

(c) Set $X = I_1 + \cdots + I_n$, and $Y = J_1 + \cdots + J_n$ and use (a) and (b) to compute $E(XY)$, $\operatorname{Var} X$, and $\operatorname{Var} Y$.

(d) Use (c) to compute $\rho(X, Y)$. Compare with the corresponding correlation coefficient in Exercise 24 with $p_1 = r_1/r$ and $p_2 = r_2/r$.

26 Let X be a random variable having density f given by

$$f(x) = \begin{cases} 1/18, & x = 1, 3, \\ 16/18, & x = 2. \end{cases}$$

Show that there is a value of δ such that $P(|X - \mu| \geq \delta) = \text{Var } X/\delta^2$, so that in general the bound given by Chebyshev's inequality cannot be improved.

27 A bolt manufacturer knows that 5% of his production is defective. He gives a guarantee on his shipment of 10,000 parts by promising to refund the money if more than a bolts are defective. How small can the manufacturer choose a and still be assured that he need not give a refund more than 1% of the time?

28 Let X have a Poisson density with parameter λ. Use Chebyshev's inequality to verify the following inequalities:

$$\text{(a) } P\left(X \leq \frac{\lambda}{2}\right) \leq \frac{4}{\lambda}; \qquad \text{(b) } P(X \geq 2\lambda) \leq \frac{1}{\lambda}.$$

29 Let X be a nonnegative integer-valued random variable whose probability generating function $\Phi_X(t) = Et^X$ is finite for all t and let x_0 be a positive number. By arguing as in the proof of Chebyshev's inequality, verify the following inequalities:

$$\text{(a) } P(X \leq x_0) \leq \frac{\Phi_X(t)}{t^{x_0}}, \qquad 0 \leq t \leq 1;$$

$$\text{(b) } P(X \geq x_0) \leq \frac{\Phi_X(t)}{t^{x_0}}, \qquad t \geq 1.$$

30 Let X have a Poisson density with parameter λ. Verify the following inequalities:

$$\text{(a) } P\left(X \leq \frac{\lambda}{2}\right) \leq \left(\frac{2}{e}\right)^{\lambda/2}; \qquad \text{(b) } P(X \geq 2\lambda) \leq \left(\frac{e}{4}\right)^{\lambda}.$$

Hint: Use calculus to minimize the right sides of the inequalities in Exercise 29. These inequalities are much sharper, especially for large values of λ, than are those given in Exercise 28.

Exercises 31–36 develop and apply the notions of conditional density and conditional expectation.

Let X and Y be discrete random variables. The *conditional density* $f_{Y|X}(y \mid x)$ *of Y given $X = x$ is defined by*

$$f_{Y|X}(y \mid x) = \begin{cases} P(Y = y \mid X = x), & \text{if } P(X = x) > 0, \\ 0, & \text{elsewhere.} \end{cases}$$

For any x such that $P(X = x) > 0$ it follows that $f_{Y|X}(y \mid x)$ is a density in y. Example 14(d) of Chapter 3 can be interpreted as saying that if X and Y are independent and geometrically distributed with parameter p, then, for $z \geq 0$, the conditional density of Y given $X + Y = z$ is the uniform density on $\{0, 1, \ldots, z\}$.

Let Y have finite expectation. The *conditional expectation of Y given $X = x$* is defined as the mean of the conditional density of Y given $X = x$, i.e., as

$$E[Y \mid X = x] = \sum_y y f_{Y|X}(y \mid x).$$

31 Verify the following properties of the conditional density and conditional expectation:

(a) $f_Y(y) = \sum_x f_X(x) f_{Y|X}(y \mid x)$; (b) $EY = \sum_x f_X(x) E[Y \mid X = x]$.

32 Let X and Y be independent random variables each having a geometric density with parameter p. Find $E[Y \mid X + Y = z]$ where z is a nonnegative integer. *Hint:* Use Example 14(d) and Exercise 8.

33 Let X and Y be two independent Poisson distributed random variables having parameters λ_1 and λ_2 respectively. Compute $E[Y \mid X + Y = z]$ where z is a nonnegative integer. *Hint:* Use the result of Exercise 35 of Chapter 3.

34 Let N be a nonnegative integer-valued random variable. Let $\{Y_n\}$, $n \geq 0$, be random variables each of which has finite expectation and is independent of N. Show that

$$E[Y_N \mid N = n] = EY_n.$$

35 Let $\{X_n\}$, $n \geq 1$, be independent random variables having a common finite mean μ and variance σ^2. Set $S_0 = 0$ and $S_n = X_1 + \cdots + X_n$, $n \geq 1$. Let N be a nonnegative integer-valued random variable having finite mean and variance, and suppose that N is independent of all random variables defined in terms of $\{X_n\}$, $n \geq 1$. Then S_N has finite mean and variance. Show that

$$ES_N = \mu EN, \qquad ES_N^2 = \sigma^2 EN + \mu^2 EN^2,$$

and

$$\operatorname{Var} S_N = \sigma^2 EN + \mu^2 \operatorname{Var} N.$$

Hint: Use Exercises 31(b) and 34.

36 Obtain the results of Exercise 35 by differentiating the probability generating function of S_N, found in Example 19 of Chapter 3, and setting $t = 1$.

5 Continuous Random Variables

In Chapter 3 we considered discrete random variables and their densities, e.g., binomial, hypergeometric, and Poisson. In applications, these random variables typically denote the number of objects of a certain type, such as the number of red balls drawn in a random sample of size n with or without replacement or the number of calls into a telephone exchange in one minute.

There are many situations, both theoretical and applied, in which the natural random variables to consider are "continuous" rather than discrete. Tentatively we can define a continuous random variable X on a probability space Ω as a function $X(\omega)$, $\omega \in \Omega$, such that

$$P(\{\omega \mid X(\omega) = x\}) = 0, \qquad -\infty < x < \infty,$$

that is, such that X takes on any specific value x with probability zero.

It is easy to think of examples of continuous random variables. As a first illustration, consider a probabilistic model for the decay times of a finite number of radioactive particles. Let T be the random variable denoting the time until the first particle decays. Then T would be a continuous random variable, for the probability is zero that the first decay occurs exactly at any specific time (e.g., $T = 2.0000\ldots$ seconds). As a second illustration, consider the experiment of choosing a point at random from a subset S of Euclidean n-space having finite nonzero n-dimensional volume (recall the discussion of this in Chapter 1). Let X be the random variable denoting the first coordinate of the point chosen. It is clear that X will take on any specific value with probability zero. Suppose, for example, that $n = 2$ and S is a disk in the plane centered at the origin and having unit radius. Then the set of points in S having first coordinate zero is a line segment in the plane. Any such line segment has *area* zero and hence probability zero.

Generally speaking, random variables denoting measurements of such physical quantities as spatial coordinates, weight, time, temperature, and voltage are most conveniently described as continuous random variables. Random variables which count objects or events are clear examples of discrete random variables.

There are cases, however, in which either discrete or continuous formulations could be appropriate. Thus, although we would normally consider measurement of length as a continuous random variable, we could consider the measurement as being rounded off to a certain number of decimal places and therefore as being a discrete random variable.

5.1. Random variables and their distribution functions

In applications, a random variable denotes a numerical quantity defined in terms of the outcome of a random experiment. Mathematically, however, a random variable X is a real-valued function defined on a probability space. Naturally, we want $P(X \leq x)$ to be well defined for every real number x. In other words, if $(\Omega, \mathscr{A}, P)$ is the probability space on which X is defined, we want

$$\{\omega \mid X(\omega) \leq x\}$$

to be an event (i.e., a member of $\mathscr{A}$). This leads to the following definitions.

Definition 1 *A random variable X on a probability space $(\Omega, \mathscr{A}, P)$ is a real-valued function $X(\omega)$, $\omega \in \Omega$, such that for $-\infty < x < \infty$, $\{\omega \mid X(\omega) \leq x\}$ is an event.*

Definition 2 *The distribution function F of a random variable X is the function*

$$F(x) = P(X \leq x), \qquad -\infty < x < \infty.$$

The distribution function is useful in computing various probabilities associated with the random variable X. An example of this is the formula

$$(1) \qquad P(a < X \leq b) = F(b) - F(a), \qquad a \leq b.$$

In order to verify (1), set $A = \{\omega \mid X(\omega) \leq a\}$ and $B = \{\omega \mid X(\omega) \leq b\}$. Then $A \subset B$ and, by the definition of a random variable, both A and B are events. Hence $\{\omega \mid a < X \leq b\} = B \cap A^c$ is an event and (1) is a special case of the fact proven in Section 1.3 that if $A \subset B$, then

$$P(B \cap A^c) = P(B) - P(A).$$

Example 1. Consider the experiment of choosing a point at random from the disk of radius R in the plane centered at the origin. To make the experiment more interesting, we can think of it as the result of throwing a dart at a disk-shaped target. Associated with this experiment is the uniform probability space described in Section 1.2. Let X be the random variable denoting the distance of the point chosen from the origin. The distribution function of X is easily computed. If $0 \leq x \leq R$, the event $\{\omega \mid X(\omega) \leq x\}$ is the disk in the plane, of radius x centered at the origin. Its area is πx^2. Thus by the definition of a uniform probability space,

$$P(X \leq x) = \frac{\pi x^2}{\pi R^2} = \frac{x^2}{R^2}, \qquad 0 \leq x \leq R.$$

If $x < 0$, then $P(X \leq x) = 0$. If $x > R$, then $P(X \leq x) = 1$. Thus the distribution function F of the random variable X is given by

$$(2) \qquad F(x) = \begin{cases} 0, & x < 0, \\ x^2/R^2, & 0 \leq x \leq R, \\ 1, & x > R. \end{cases}$$

The graph of F is given in Figure 1. It follows from Formulas (1) and (2) that if $0 \leq a \leq b \leq R$, then

$$P(a < X \leq b) = F(b) - F(a) = \frac{b^2 - a^2}{R^2}.$$

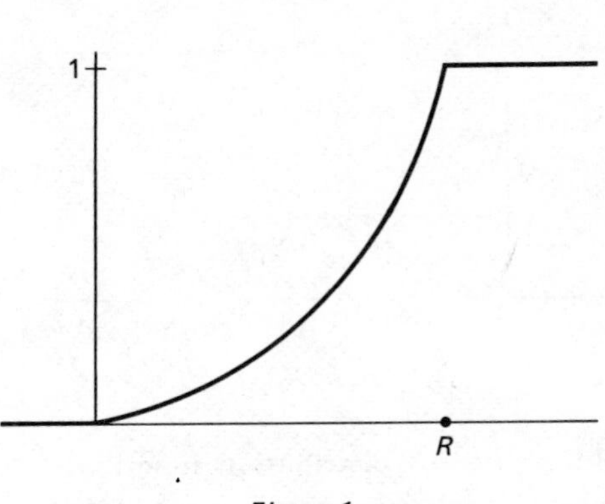

Figure 1

Example 2. Consider a probability model for the decay times of a finite number of radioactive particles. Let X denote the time to decay for a specific particle. Find the distribution function of X.

As we saw in Section 1.1, for a suitable positive value for λ,

$$P(a < X \leq b) = e^{-\lambda a} - e^{-\lambda b}, \qquad 0 \leq a \leq b < \infty.$$

Since X takes on only positive values, $P(X \leq x) = 0$ for $x \leq 0$ and, in particular, $P(X \leq 0) = 0$. For $0 < x < \infty$,

$$\begin{aligned} P(X \leq x) &= P(X \leq 0) + P(0 < X \leq x) \\ &= P(0 < X \leq x) \\ &= 1 - e^{-\lambda x}. \end{aligned}$$

Thus X has the distribution function F given by

$$(3) \qquad F(x) = \begin{cases} 0, & x \leq 0, \\ 1 - e^{-\lambda x}, & x > 0. \end{cases}$$

Of course, discrete random variables also have distribution functions, two of which were computed in Examples 10 and 11 of Chapter 3.

Example 3. Let X have a binomial distribution with parameters $n = 2$ and $p = 1/2$. Then $f(0) = 1/4$, $f(1) = 1/2$, and $f(2) = 1/4$. Consequently

$$F(x) = \begin{cases} 0, & x < 0, \\ 1/4, & 0 \le x < 1, \\ 3/4, & 1 \le x < 2, \\ 1, & 2 \le x. \end{cases}$$

The graph of this distribution function is given in Figure 2.

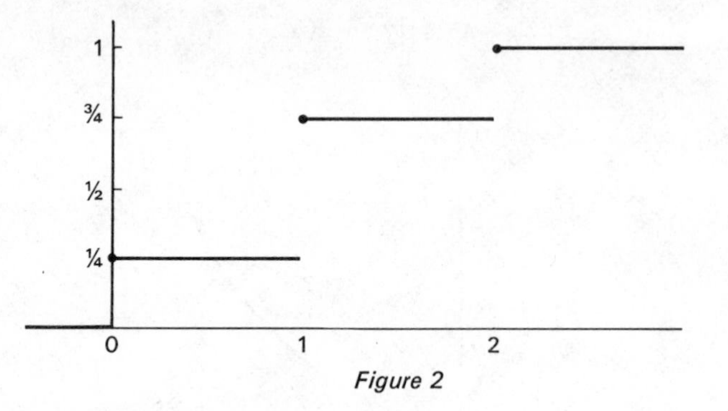

Figure 2

5.1.1. Properties of distribution functions. Not all functions can arise as distribution functions, for the latter must satisfy certain conditions. Let X be a random variable and let F be its distribution function. Then

(i) $0 \le F(x) \le 1$ *for all* x.

(ii) F *is a nondecreasing function of* x.

Property (i) follows immediately from the defining property $F(x) = P(X \le x)$. To see that (ii) holds we need only note that if $x < y$, then

$$F(y) - F(x) = P(x < X \le y) \ge 0.$$

A function f is said to have a right-hand (left-hand) limit L at x if $f(x + h) \to L$ as $h \to 0$ when h is restricted to positive (negative) values. The right-hand and left-hand limits, when they exist, are denoted respectively by $f(x+)$ and $f(x-)$. It is not hard to show that if f is bounded and either nondecreasing or nonincreasing, then $f(x+)$ and $f(x-)$ exist for all x. Under the same conditions, f has limits $f(-\infty)$ as $x \to -\infty$ and $f(+\infty)$ as $x \to +\infty$.

From properties (i) and (ii) and the discussion in the preceding paragraph, it follows that the distribution function F has limits $F(x+)$ and $F(x-)$ for all x as well as the limits $F(-\infty)$ and $F(+\infty)$.

(iii) $F(-\infty) = 0$ *and* $F(+\infty) = 1$.

(iv) $F(x+) = F(x)$ *for all* x.

In order to evaluate $F(-\infty)$ and $F(+\infty)$ we need only find the limits of $F(n)$ as $n \to -\infty$ and $n \to +\infty$. (This is because F is nondecreasing.) Set

$$B_n = \{\omega \mid X(\omega) \le n\}.$$

Then $\cdots \subseteq B_{-2} \subseteq B_{-1} \subseteq B_0 \subseteq B_1 \subseteq B_2 \subseteq \cdots$. Also

$$\bigcap_{n=0}^{-\infty} B_n = \varnothing \quad \text{and} \quad \bigcup_{n=0}^{+\infty} B_n = \Omega.$$

It now follows from the results of Theorem 1 of Chapter 1 that

$$\lim_{n \to -\infty} P(B_n) = P(\varnothing) = 0 \quad \text{and} \quad \lim_{n \to +\infty} P(B_n) = P(\Omega) = 1.$$

Since $F(n) = P(X \le n) = P(B_n)$, we have that

$$F(-\infty) = \lim_{n \to -\infty} F(n) = \lim_{n \to -\infty} P(B_n) = 0$$

and similarly that $F(+\infty) = 1$.

Property (iv) states that F is a right-continuous function and

$$F(x+) = P(X \le x), \qquad -\infty < x < \infty. \tag{4}$$

A closely related result is

$$F(x-) = P(X < x), \qquad -\infty < x < \infty. \tag{5}$$

The proofs of (4) and (5) are similar to the proof of (iii). To prove (4), for example, we need only show that $F(x + 1/n) \to P(X \le x)$ as $n \to +\infty$. This can be done by setting

$$B_n = \left\{\omega \mid X(\omega) \le x + \frac{1}{n}\right\},$$

noting that $\bigcap_n B_n = \{\omega \mid X(\omega) \le x\}$ and repeating the argument of (iii).

From (4) and (5) we see immediately that

$$F(x+) - F(x-) = P(X = x), \qquad -\infty < x < \infty. \tag{6}$$

This formula states that if $P(X = x) > 0$, then F has a jump at x of magnitude $P(X = x)$. If $P(X = x) = 0$, then F is continuous at x. We recall from the introduction to this chapter the concept of a continuous random variable.

Definition 3 *A random variable X is called a continuous random variable if*

$$P(X = x) = 0, \qquad -\infty < x < \infty.$$

We now see that X is a continuous random variable if and only if its distribution function F is continuous at every x, that is, F is a continuous function. If X is a continuous random variable, then in addition to (1) we have that

$$(7) \qquad P(a < X < b) = P(a \le X \le b) = P(a \le X < b) = F(b) - F(a),$$

so that $<$ and $\le$ can be used indiscriminately in this context. The various properties of a distribution function are illustrated in Figure 3. (Note that the random variable having this distribution function would be neither discrete nor continuous.)

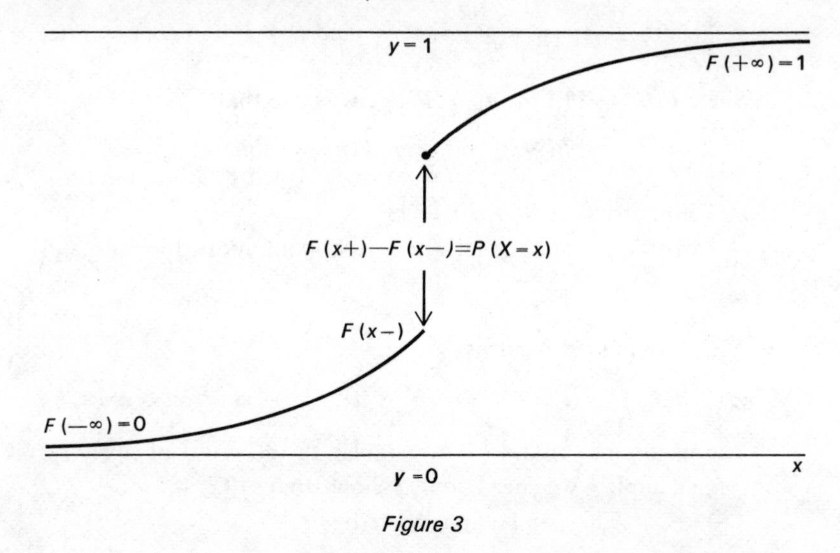

Figure 3

Consider the random variable X defined in Example 1. From Formula (2) or Figure 1 we see that its distribution function is continuous. Thus X is a continuous random variable. Similarly it is clear from (3) that the random variable from Example 2 is a continuous random variable.

Most random variables arising in practical applications are either discrete or continuous. There are some exceptions. Consider Example 2. In this example X represents the time to decay of a specific particle. If the experiment lasts only a specified time, say until time $t_0 > 0$, and the particle has not decayed by this time, then its true decay time X will not be observed. One possible way out of this difficulty is to define a new random variable Y as follows

$$Y(\omega) = \begin{cases} X(\omega) & \text{if } X(\omega) \le t_0, \\ t_0 & \text{if } X(\omega) > t_0. \end{cases}$$

Thus Y is the decay time, if this time is observed (i.e., is less than or equal to t_0), and otherwise $Y = t_0$. The distribution function F_Y of Y is given by

$$F_Y(y) = \begin{cases} 0, & y < 0, \\ 1 - e^{-\lambda y}, & 0 \leq y < t_0, \\ 1, & y \geq t_0. \end{cases}$$

The distribution function has a jump at $y = t_0$ of magnitude $e^{-\lambda t_0}$. Thus it is clear that the random variable Y we have constructed is neither discrete nor continuous.

We have defined distribution functions in terms of random variables. They can be defined directly.

Definition 4 *A distribution function is any function F satisfying properties* (i)–(iv); *that is,*

(i) $0 \leq F(x) \leq 1$ *for all x,*

(ii) *F is a nondecreasing function of x,*

(iii) $F(-\infty) = 0$ *and* $F(+\infty) = 1$,

(iv) $F(x+) = F(x)$ *for all x.*

In more advanced books it is shown that *if F is a distribution function, there is necessarily a probability space and a random variable X defined on that space such that F is the distribution function of X.*

5.2. Densities of continuous random variables

In practice, continuous distribution functions are usually defined in terms of density functions.

Definition 5 *A density function* (*with respect to integration*) *is a nonnegative function f such that*

$$\int_{-\infty}^{\infty} f(x)\, dx = 1.$$

Note that if f is a density function, then the function F defined by

$$F(x) = \int_{-\infty}^{x} f(y)\, dy, \qquad -\infty < x < \infty, \tag{8}$$

is a continuous function satisfying properties (i)–(iv) of Section 5.1.1. Thus (8) defines a continuous distribution function. We say that this distribution function has density f. It is possible but difficult to construct examples of continuous distribution functions that do not have densities. Those that do have densities are called *absolutely continuous* distribution functions.

If X is a continuous random variable having F as its distribution function, where F is given by (8), then f is also called the density of X. In the sequel we will use "density function" to refer to either discrete density functions or density functions with respect to integration. It should be clear from the context which type of density function is under consideration. For example, the phrase "let X be a continuous random variable having density f" necessarily implies that f is a density function with respect to integration.

It follows from (1) and (8) that if X is a continuous random variable having density f, then

$$(9) \qquad P(a \leq X \leq b) = \int_a^b f(x)\, dx, \qquad a \leq b,$$

or somewhat more generally, that

$$(10) \qquad P(X \in A) = \int_A f(x)\, dx$$

if A is a finite or countably infinite union of disjoint intervals. Thus $P(X \in A)$ can be represented as the area under the curve f as x ranges over the set A (see Figure 4).

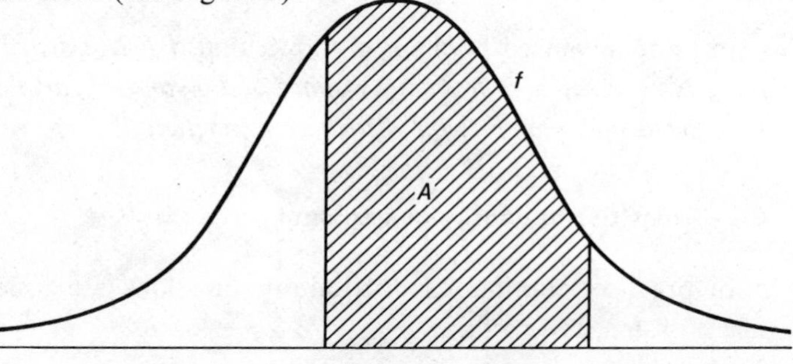

Figure 4

In most applications, the easiest way to compute densities of continuous random variables is to differentiate (8) and obtain

$$(11) \qquad f(x) = F'(x), \qquad -\infty < x < \infty.$$

Strictly speaking, (11) holds at all points x where f is continuous.

Example 4. Let X be the random variable from Example 1 having the distribution function F given by (2). Then

$$(12) \qquad F'(x) = \begin{cases} 0, & x < 0, \\ 2x/R^2, & 0 \leq x < R, \\ 0, & x > R. \end{cases}$$

At $x = R$ the function F is not differentiable. If, however, we define f by $f(x) = F'(x)$, $x \neq R$, and $f(R) = 0$, then this f will be a density for F.

We note that (8) does not define f uniquely since we can always change the value of a function at a finite number of points without changing the integral of the function over intervals. One typical way to define f is by setting $f(x) = F'(x)$ whenever $F'(x)$ exists and $f(x) = 0$ otherwise. This defines a density of F provided that F is everywhere continuous and that F' exists and is continuous at all but a finite number of points.

There are other ways to derive or verify formulas for the density of a continuous distribution function F. Given a density function f we can show that f is a density function of F by verifying that (8) holds. Alternatively, we can reverse this process and show that F can be written in the form (8) for some nonnegative function f. Then f is necessarily a density function of F. These methods, essentially equivalent to each other, are usually more complicated than is differentiation. However, they are rigorous and avoid special consideration of points where $F'(x)$ fails to exist.

We will illustrate these methods in our first example of the following subsection.

5.2.1. Change of variable formulas.

Let X be a continuous random variable having density f. We will discuss methods for finding the density of a random variable Y which is a function of X.

Example 5. Let X be a continuous random variable having density f. Find the density of the random variable $Y = X^2$.

To solve this problem we first let F and G denote the respective distribution functions of X and Y. Then $G(y) = 0$ for $y \leq 0$. For $y > 0$

$$\begin{aligned} G(y) &= P(Y \leq y) = P(X^2 \leq y) \\ &= P(-\sqrt{y} \leq X \leq \sqrt{y}) \\ &= F(\sqrt{y}) - F(-\sqrt{y}) \end{aligned}$$

and by differentiation we see that

$$\begin{aligned} G'(y) &= \frac{1}{2\sqrt{y}} (F'(\sqrt{y}) + F'(-\sqrt{y})) \\ &= \frac{1}{2\sqrt{y}}(f(\sqrt{y}) + f(-\sqrt{y})). \end{aligned}$$

Thus $Y = X^2$ has density g given by

$$g(y) = \begin{cases} \dfrac{1}{2\sqrt{y}} (f(\sqrt{y}) + f(-\sqrt{y})) & \text{for} \quad y > 0, \\ 0 & \text{for} \quad y \leq 0. \end{cases} \tag{13}$$

Although (13) is valid in general, our derivation depended on differentiation, which may not be valid at all points. To give an elementary but completely rigorous proof of (13), we can define g by the right side of (13) and write for $x > 0$

$$\int_{-\infty}^{x} g(y)\,dy = \int_{0}^{x} \frac{1}{2\sqrt{y}}(f(\sqrt{y}) + f(-\sqrt{y}))\,dy.$$

By making the change of variable $z = \sqrt{y}$ (so that $dz = dy/2\sqrt{y}$), we obtain

$$\begin{aligned}\int_{-\infty}^{x} g(y)\,dy &= \int_{0}^{\sqrt{x}} (f(z) + f(-z))\,dz \\ &= \int_{-\sqrt{x}}^{\sqrt{x}} f(z)\,dz \\ &= F(\sqrt{x}) - F(-\sqrt{x}) = G(x),\end{aligned}$$

so that g is indeed a density of G.

Hereafter we will freely use differentiation to establish formulas such as (13), knowing that we could if necessary provide alternative derivations via integration.

Let us now use (13) to find the density of X^2, where X is the random variable defined in Example 1. The density of X was found in Example 4 to be $f(x) = 2x/R^2$ for $0 \le x < R$ and $f(x) = 0$ elsewhere. Thus by (13), X^2 has density g given by

$$g(y) = \frac{1}{2\sqrt{y}} \frac{2\sqrt{y}}{R^2} = \frac{1}{R^2}, \qquad 0 < y < R^2,$$

and $g(y) = 0$ elsewhere. This density is a uniform density on $(0, R^2)$ according to the following.

Definition 6 *Let a and b be constants with $a < b$. The uniform density on the interval (a, b) is the density f defined by*

$$f(x) = \begin{cases} (b - a)^{-1} & \text{for } a < x < b, \\ 0 & \text{elsewhere.} \end{cases} \tag{14}$$

The distribution function corresponding to (14) is given by

$$F(x) = \begin{cases} 0, & x < a, \\ (x - a)/(b - a), & a \le x \le b, \\ 1, & x > b. \end{cases} \tag{15}$$

It is not difficult to find other examples of uniformly distributed random variables. If a well-balanced dial is spun around and comes to rest after a large number of revolutions, it is reasonable to assume that the angle of the dial after it stops moving (suitably defined in radians) is uniformly distributed on $(-\pi, \pi)$ or, equivalently, on $(0, 2\pi)$. In applications of probability theory to numerical analysis, it is often assumed that the rounding error caused by dropping all digits more than n places beyond the decimal point is uniformly distributed on $(0, 10^{-n})$.

Example 6. Let X be uniformly distributed on $(0, 1)$. Find the density of $Y = -\lambda^{-1} \log (1 - X)$ for $\lambda > 0$.

Let G denote the distribution function of Y. We observe first that Y is a positive random variable and consequently $G(y) = 0$ for $y \leq 0$. For $y > 0$ we have

$$\begin{aligned} G(y) = P(Y \leq y) &= P(-\lambda^{-1} \log (1 - X) \leq y) \\ &= P(\log (1 - X) \geq -\lambda y) \\ &= P(1 - X \geq e^{-\lambda y}) \\ &= P(X \leq 1 - e^{-\lambda y}) \\ &= 1 - e^{-\lambda y}. \end{aligned}$$

Hence $G'(y) = \lambda e^{-\lambda y}$ for $y > 0$ and $G'(y) = 0$ for $y < 0$. The density of Y is therefore given by

$$g(y) = \begin{cases} \lambda e^{-\lambda y}, & y > 0, \\ 0, & y \leq 0. \end{cases} \tag{16}$$

This density is called the *exponential density with parameter* λ and will be discussed further in the next section.

The above example is a special case of problems that can be solved by means of the following theorem.

Theorem 1 *Let φ be a differentiable strictly increasing or strictly decreasing function on an interval I, and let $\varphi(I)$ denote the range of φ and φ^{-1} the inverse function to φ. Let X be a continuous random variable having density f such that $f(x) = 0$ for $x \notin I$. Then $Y = \varphi(X)$ has density g given by $g(y) = 0$ for $y \notin \varphi(I)$ and*

$$g(y) = f(\varphi^{-1}(y)) \left| \frac{d}{dy} \varphi^{-1}(y) \right|, \qquad y \in \varphi(I). \tag{17}$$

It is somewhat more suggestive to write (17) in the equivalent form

$$g(y) = f(x) \left| \frac{dx}{dy} \right|, \qquad y \in \varphi(I) \quad \text{and} \quad x = \varphi^{-1}(y) \tag{18}$$

(or alternatively $g(y)|dy| = f(x)|dx|$).

In order to derive (17), let F and G denote the respective distribution functions of X and Y. Suppose first that φ is strictly increasing (i.e.,

$\varphi(x_1) < \varphi(x_2)$ if $x_1 < x_2$, $x_1 \in I$ and $x_2 \in I$). Then φ^{-1} is strictly increasing on $\varphi(I)$ and for $y \in \varphi(I)$,

$$\begin{aligned} G(y) &= P(Y \le y) \\ &= P(\varphi(X) \le y) \\ &= P(X \le \varphi^{-1}(y)) \\ &= F(\varphi^{-1}(y)). \end{aligned}$$

Thus by the chain rule for differentiation,

$$\begin{aligned} G'(y) &= \frac{d}{dy} F(\varphi^{-1}(y)) \\ &= F'(\varphi^{-1}(y)) \frac{d}{dy} \varphi^{-1}(y) \\ &= f(\varphi^{-1}(y)) \frac{d}{dy} \varphi^{-1}(y). \end{aligned}$$

Now

$$\frac{d}{dy} \varphi^{-1}(y) = \left| \frac{d}{dy} \varphi^{-1}(y) \right|$$

because φ^{-1} is strictly increasing so that (17) holds. Suppose next that φ is strictly decreasing on I. Then φ^{-1} is strictly decreasing on $\varphi(I)$, and for $y \in \varphi(I)$

$$\begin{aligned} G(y) &= P(Y \le y) \\ &= P(\varphi(X) \le y) \\ &= P(X \ge \varphi^{-1}(y)) \\ &= 1 - F(\varphi^{-1}(y)). \end{aligned}$$

Thus

$$\begin{aligned} G'(y) &= -F'(\varphi^{-1}(y)) \frac{d}{dy} \varphi^{-1}(y) \\ &= f(\varphi^{-1}(y)) \left(-\frac{d}{dy} \varphi^{-1}(y) \right). \end{aligned}$$

Now

$$-\frac{d}{dy} \varphi^{-1}(y) = \left| \frac{d}{dy} \varphi^{-1}(y) \right|$$

because φ^{-1} is strictly decreasing. Therefore in either case we see that G has the density g given by (17). ∎

Example 7. Let X be a random variable having an exponential density with parameter λ. Find the density of $Y = X^{1/\beta}$, where $\beta \neq 0$.

According to the definition given in the previous example, X has the density f given by $f(x) = \lambda e^{-\lambda x}$ for $x > 0$ and $f(x) = 0$ for $x \le 0$. The above theorem is applicable with $\varphi(x) = x^{1/\beta}$, $x > 0$. The equation $y = x^{1/\beta}$ has solution $x = y^\beta$ which yields $dx/dy = \beta y^{\beta-1}$. Thus by (18), Y has density g given by

$$g(y) = \begin{cases} |\beta|\lambda y^{\beta-1}e^{-\lambda y^\beta}, & y > 0, \\ 0, & y \le 0. \end{cases}$$

Example 8. Let X be a continuous random variable having density f and let a and b be constants such that $b \ne 0$. Then by Theorem 1, the random variable $Y = a + bX$ has density given by

$$(19) \qquad g(y) = \frac{1}{|b|} f\left(\frac{y-a}{b}\right), \qquad -\infty < y < \infty.$$

As an illustration of this formula, let X be the random variable defined in Example 1. In Example 4 we found its density function f to be given by $f(x) = 2x/R^2$ for $0 < x < R$ and $f(x) = 0$ elsewhere. Consider the random variable $Y = X/R$ and let g denote its density. Then by Formula (19) with $a = 0$ and $b = 1/R$,

$$g(y) = Rf(Ry) = 2y, \qquad 0 < y < 1,$$

and $g(y) = 0$ elsewhere.

The reader may prefer to derive formulas such as those of Examples 7 and 8 by using the direct method of Example 6 instead of Theorem 1.

As we have seen in the above examples, we can construct density functions by considering functions of random variables. There is another simple way of constructing density functions. Let g be any nonnegative function such that

$$0 < \int_{-\infty}^{\infty} g(x)\,dx < \infty.$$

Then g can always be normalized to yield a density function $f = c^{-1}g$, where c is the constant

$$c = \int_{-\infty}^{\infty} g(x)\,dx.$$

The following examples illustrate this method.

Example 9. Let $g(x) = x(1 - x)$, $0 \le x \le 1$, and $g(x) = 0$ elsewhere. Then

$$c = \int_0^1 x(1-x)\,dx = \left(\frac{x^2}{2} - \frac{x^3}{3}\right)\Bigg|_0^1 = \frac{1}{6}$$

and $f = c^{-1}g$ is given by $f(x) = 6x(1 - x)$, $0 \le x \le 1$, and $f(x) = 0$ elsewhere. The corresponding distribution function is given by $F(x) = 0$ for $x < 0$, $F(x) = 3x^2 - 2x^3$ for $0 \le x \le 1$, and $F(x) = 1$ for $x > 1$.

Example 10. Let $g(x) = 1/(1 + x^2)$, $-\infty < x < \infty$. From calculus we know that the indefinite integral of $1/(1 + x^2)$ is arctan x. Thus

$$c = \int_{-\infty}^{\infty} \frac{dx}{1 + x^2} = \arctan x \Big|_{-\infty}^{\infty} = \frac{\pi}{2} - \left(-\frac{\pi}{2}\right) = \pi.$$

Consequently $f = c^{-1}g$ is given by

$$f(x) = \frac{1}{\pi(1 + x^2)}, \qquad -\infty < x < \infty.$$

This density is known as the *Cauchy density*. The corresponding distribution function is given by

$$F(x) = \frac{1}{2} + \frac{1}{\pi} \arctan x, \qquad -\infty < x < \infty.$$

For an illustration of a Cauchy distributed random variable we have the following:

Example 11. Let X denote the tangent of an angle (measured in radians) chosen at random from $(-\pi/2, \pi/2)$. Find the distribution of X.

In solving this problem we will let Θ be the random variable denoting the angle chosen measured in radians. Now $X = \tan \Theta$ and hence (see Figure 5) for $-\infty < x < \infty$,

$$\begin{aligned} P(X \le x) &= P(\tan \Theta \le x) \\ &= P\left(-\frac{\pi}{2} < \Theta \le \arctan x\right) \\ &= \frac{1}{\pi}\left(\arctan x - \left(-\frac{\pi}{2}\right)\right) \\ &= \frac{1}{2} + \frac{1}{\pi} \arctan x. \end{aligned}$$

Thus X has the Cauchy distribution.

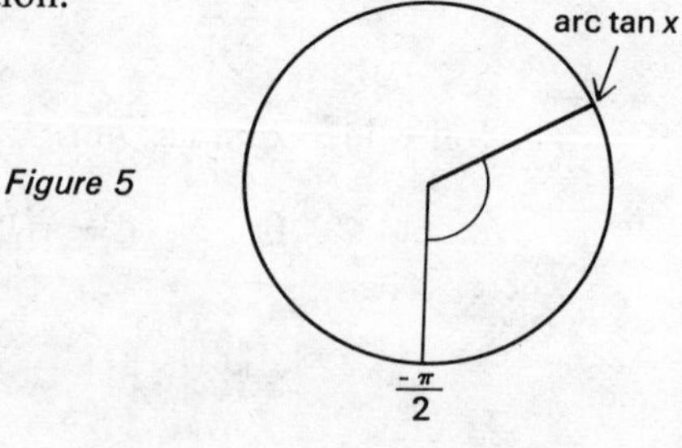

Figure 5

5.2.2. Symmetric densities. We will close this section by discussing symmetric densities and symmetric random variables. A density function f is called symmetric if $f(-x) = f(x)$ for all x. The Cauchy density and the uniform density on $(-a, a)$ are both symmetric. A random variable X is said to be symmetric if X and $-X$ have the same distribution function. The next result shows that these two concepts of symmetry are very closely related.

Theorem 2 *Let X be a random variable that has a density. Then f has a symmetric density if and only if X is a symmetric random variable.*

Proof. We will prove this result for continuous random variables. The proof for discrete random variables is similar. In our proof we will use the fact that for any integrable function f

$$\int_{-\infty}^{x} f(-y)\,dy = \int_{-x}^{\infty} f(y)\,dy, \qquad -\infty < x < \infty.$$

Suppose first that X has a symmetric density f. Then

$$\begin{aligned} P(-X \le x) &= P(X \ge -x) \\ &= \int_{-x}^{\infty} f(y)\,dy \\ &= \int_{-\infty}^{x} f(-y)\,dy \\ &= \int_{-\infty}^{x} f(y)\,dy \\ &= P(X \le x), \end{aligned}$$

so that X and $-X$ have the same distribution function.

Suppose conversely that X and $-X$ have a common density g. Define f by $f(x) = (g(x) + g(-x))/2$. Then f is clearly a symmetric density function. Also

$$\begin{aligned} \int_{-\infty}^{x} f(y)\,dy &= 1/2 \int_{-\infty}^{x} g(y)\,dy + 1/2 \int_{-\infty}^{x} g(-y)\,dy \\ &= 1/2 \int_{-\infty}^{x} g(y)\,dy + 1/2 \int_{-x}^{\infty} g(y)\,dy \\ &= 1/2[P(X \le x)] + 1/2[P(-X \ge -x)] \\ &= P(X \le x). \end{aligned}$$

Thus X has the symmetric density f, as desired. ∎

If a continuous distribution function F has a symmetric density f, then $F(0) = 1/2$. The values of F for negative x's can be calculated from the values of F for positive x's. For

$$\begin{aligned} F(-x) &= \int_{-\infty}^{-x} f(y)\,dy \\ &= \int_{x}^{\infty} f(-y)\,dy \\ &= \int_{x}^{\infty} f(y)\,dy \\ &= \int_{-\infty}^{\infty} f(y)\,dy - \int_{-\infty}^{x} f(y)\,dy \end{aligned}$$

and hence

$$F(-x) = 1 - F(x), \qquad -\infty < x < \infty. \tag{20}$$

For this reason, when tables of such a distribution function are constructed, usually only nonnegative values of x are presented.

5.3. Normal, exponential, and gamma densities

In this section we will discuss three of the most important families of density functions in probability theory and statistics.

5.3.1. Normal densities. Let $g(x) = e^{-x^2/2}$, $-\infty < x < \infty$. In order to normalize g to make it a density we need to evaluate the constant

$$c = \int_{-\infty}^{\infty} e^{-x^2/2}\,dx.$$

There is no simple formula for the indefinite integral of $e^{-x^2/2}$. The easiest way to evaluate c is by a very special trick in which we write c as a two-dimensional integral and introduce polar coordinates. To be specific

$$\begin{aligned} c^2 &= \int_{-\infty}^{\infty} e^{-x^2/2}\,dx \int_{-\infty}^{\infty} e^{-y^2/2}\,dy \\ &= \int_{-\infty}^{\infty}\int_{-\infty}^{\infty} e^{-(x^2+y^2)/2}\,dx\,dy \\ &= \int_0^{\infty}\left(\int_{-\pi}^{\pi} e^{-r^2/2} r\,d\theta\right) dr \\ &= 2\pi \int_0^{\infty} r e^{-r^2/2}\,dr \\ &= -2\pi e^{-r^2/2}\Big|_0^{\infty} \\ &= 2\pi. \end{aligned}$$

Thus $c = \sqrt{2\pi}$ and the normalized form of g is given by

$$f(x) = (2\pi)^{-1/2} e^{-x^2/2}, \qquad -\infty < x < \infty.$$

We also record the formula

$$\int_{-\infty}^{\infty} e^{-x^2/2}\, dx = \sqrt{2\pi}. \tag{21}$$

The density just derived is called the *standard normal density* and is usually denoted by φ, so that

$$\varphi(x) = \frac{1}{\sqrt{2\pi}} e^{-x^2/2}, \qquad -\infty < x < \infty. \tag{22}$$

The standard normal density is clearly symmetric. The distribution function of φ is denoted by Φ. There is no simple formula for Φ so it must be evaluated numerically. Computer routines and tables such as Table I at the back of this book are available for computing Φ. Since φ is symmetric, (20) is applicable and

$$\Phi(-x) = 1 - \Phi(x), \qquad -\infty < x < \infty. \tag{23}$$

Let X be a random variable having the standard normal density φ and let $Y = \mu + \sigma X$, where $\sigma > 0$. Then by Formula (19), Y has the density g given by

$$g(y) = \frac{1}{\sigma\sqrt{2\pi}} e^{-(y-\mu)^2/2\sigma^2}, \qquad -\infty < y < \infty.$$

This density is called the normal density with mean μ and variance σ^2 and is denoted by $n(\mu, \sigma^2)$ or $n(y; \mu, \sigma^2)$, $-\infty < y < \infty$. Thus

$$n(y; \mu, \sigma^2) = \frac{1}{\sigma\sqrt{2\pi}} e^{-(y-\mu)^2/2\sigma^2} = \frac{1}{\sigma}\varphi\left(\frac{y-\mu}{\sigma}\right), \quad -\infty < y < \infty. \tag{24}$$

Since we have not yet defined moments of continuous random variables, we should temporarily think of μ and σ^2 as the two parameters of the family of normal densities. The corresponding distribution function can be calculated in terms of Φ, for

$$\begin{aligned} P(Y \le y) &= P(\mu + \sigma X \le y) \\ &= P\left(X \le \frac{y-\mu}{\sigma}\right) \\ &= \Phi\left(\frac{y-\mu}{\sigma}\right). \end{aligned}$$

It follows that if Y is distributed as $n(\mu, \sigma^2)$ and $a \le b$, then

$$P(a \le Y \le b) = \Phi\left(\frac{b-\mu}{\sigma}\right) - \Phi\left(\frac{a-\mu}{\sigma}\right). \tag{25}$$

For example, let Y be distributed as $n(1, 4)$ and let $a = 0$ and $b = 3$. We find from Table I that

$$
\begin{aligned}
P(0 \leq Y \leq 3) &= \Phi(1) - \Phi(-1/2) = \Phi(1) - (1 - \Phi(1/2)) \\
&= .8413 - .3085 \\
&= .5328.
\end{aligned}
$$

If a random variable Y is distributed as $n(\mu, \sigma^2)$, then the random variable $a + bY$, $b \neq 0$, is distributed as $n(a + b\mu, b^2\sigma^2)$. This is a direct application of (19). Alternatively, we can write $Y = \mu + \sigma X$, where X has the standard normal distribution. Then

$$a + bY = a + b(\mu + \sigma X) = (a + b\mu) + b\sigma X,$$

which is distributed as $n(a + b\mu, b^2\sigma^2)$.

Normally distributed random variables occur very often in practical applications. Maxwell's Law in physics asserts that under appropriate conditions the components of the velocity of a molecule of gas will be randomly distributed according to a normal density $n(0, \sigma^2)$, where σ^2 depends on certain physical quantities. In most applications, however, the random variables of interest will have a distribution function that is only approximately normal. For example, measurement errors in physical experiments, variability of outputs from industrial production lines, and biological variability (e.g., those of height and weight) have been found empirically to have approximately normal distributions. It has also been found, both empirically and theoretically, that random fluctuations which result from a combination of many unrelated causes, each individually insignificant, tend to be approximately normally distributed. Theoretical results in this direction are known as "central limit theorems" and have developed into one of the major research topics in probability theory. One such central limit theorem will be discussed in Chapter 7 and proved in Chapter 8. The importance of normal distributions arises also from their nice theoretical properties. An example is the property that the sum of independent normally distributed random variables is itself normally distributed. This will be proved in Chapter 6. In Volume II we will see that normal distributions also play a fundamental role in theoretical and applied statistics.

5.3.2. Exponential densities. The exponential density with parameter λ was defined in Section 5.2. It is given by

$$f(x) = \begin{cases} \lambda e^{-\lambda x}, & x \geq 0, \\ 0, & x < 0. \end{cases} \tag{26}$$

The corresponding distribution function is

$$F(x) = \begin{cases} 1 - e^{-\lambda x}, & x \geq 0, \\ 0, & x < 0. \end{cases} \tag{27}$$

From the discussion in Chapter 1 and in Example 2 of this chapter we see that exponentially distributed random variables are useful in studying decay times of radioactive particles. They are also useful in developing models involving many other waiting times, such as the time until a piece of equipment fails, the time it takes to complete a job, or the time it takes to get a new customer. Exponentially distributed random variables are also of theoretical importance, as can be seen by studying Poisson processes (see Chapter 9) or continuous time Markov chains (see Volume III).

An important property of exponentially distributed random variables is that if X is such a variable, then

$$P(X > a + b) = P(X > a)P(X > b), \qquad a \geq 0 \text{ and } b \geq 0. \tag{28}$$

(This formula is similar to the one obtained in Chapter 3 for geometrically distributed random variables.) In order to see that (28) holds, let λ denote the parameter of the exponential distribution of X. Then by (27)

$$\begin{aligned} P(X > a)P(X > b) &= e^{-\lambda a}e^{-\lambda b} \\ &= e^{-\lambda(a+b)} \\ &= P(X > a + b). \end{aligned}$$

A more suggestive but equivalent form of (28) is

$$P(X > a + b \mid X > a) = P(X > b), \qquad a \geq 0 \text{ and } b \geq 0. \tag{29}$$

Think of X as the time it takes a piece of equipment to fail after it is installed. Then (29) states that, conditioned on there having been no failure by time a, the probability of no failure in the next b units of time is equal to the unconditioned probability of no failure during the first b units of time. This implies that the aging of the piece of equipment neither increases nor decreases its probability of failing in a given length of time.

That (28) or (29) characterizes the family of exponential distributions is shown by the following result.

Theorem 3 *Let X be a random variable such that* (28) *holds. Then either $P(X > 0) = 0$ or X is exponentially distributed.*

Proof. If $P(X > 0) = 0$, then (28) holds trivially. Suppose (28) holds and $P(X > 0) \neq 0$. Then by (28) with $a = b = 0$ we see that $P(X > 0) = 1$, so that X is a positive random variable. Let F

denote the distribution function of X and define G by $G(x) = 1 - F(x) = P(X > x)$. Then G is a right-continuous, nonincreasing function, $G(0) = 1$, $G(+\infty) = 0$, and by (28)

$$G(a + b) = G(a)G(b), \qquad a > 0 \quad \text{and} \quad b > 0.$$

It follows that if $c > 0$ and m and n are positive integers, then

$$G(nc) = (G(c))^n \qquad \text{and} \qquad G(c) = (G(c/m))^m. \tag{30}$$

We claim next that $0 < G(1) < 1$. For if $G(1) = 1$, then $G(n) = (G(1))^n = 1$, which contradicts $G(+\infty) = 0$. If $G(1) = 0$, then $G(1/m) = 0$ and by right-continuity, $G(0) = 0$, another contradiction.

Since $0 < G(1) < 1$, we can write $G(1) = e^{-\lambda}$ where $0 < \lambda < \infty$. It follows from (30) that if m is a positive integer, then $G(1/m) = e^{-\lambda/m}$. A second application of (30) yields that if m and n are positive integers, then $G(n/m) = e^{-\lambda n/m}$. In other words $G(y) = e^{-\lambda y}$ holds for all positive rational numbers y. By right-continuity it follows that $G(y) = e^{-\lambda y}$ for all $y \geq 0$. This implies that $F = 1 - G$ is the exponential distribution function with parameter λ. ∎

5.3.3. Gamma densities. Before defining gamma densities in general we will first consider an example in which they arise naturally.

Example 12. Let X be a random variable having the normal density $n(0, \sigma^2)$. Find the density of the random variable $Y = X^2$.

In solving this problem we note first that the density of X is

$$f(x) = \frac{1}{\sigma\sqrt{2\pi}} e^{-x^2/2\sigma^2}, \qquad -\infty < x < \infty.$$

By Formula (13), Y has density g given by $g(y) = 0$ for $y \leq 0$ and

$$g(y) = \frac{1}{2\sqrt{y}} (f(\sqrt{y}) + f(-\sqrt{y})), \qquad y > 0.$$

This implies that

$$g(y) = \frac{1}{\sigma\sqrt{2\pi y}} e^{-y/2\sigma^2}, \qquad y > 0. \tag{31}$$

In order to define gamma densities in general, we first consider functions g of the form

$$g(x) = \begin{cases} x^{\alpha-1}e^{-\lambda x}, & x > 0, \\ 0, & x \leq 0. \end{cases}$$

Here we require $\alpha > 0$ and $\lambda > 0$ in order that g be integrable. The density in (31) corresponds to the special case $\alpha = 1/2$ and $\lambda = 1/2\sigma^2$. In normalizing g to make it a density we must evaluate

$$c = \int_0^\infty x^{\alpha-1}e^{-\lambda x}\,dx.$$

Make the change of variable $y = \lambda x$. Then

$$c = \frac{1}{\lambda^\alpha}\int_0^\infty y^{\alpha-1}e^{-y}\,dy.$$

There is no simple formula for the last integral. Instead it is used to define a function called the gamma function and denoted by Γ. Thus

$$c = \frac{1}{\lambda^\alpha}\Gamma(\alpha),$$

where

$$\Gamma(\alpha) = \int_0^\infty x^{\alpha-1}e^{-x}\,dx, \qquad \alpha > 0. \tag{32}$$

The normalized function is called the gamma density with parameters α and λ and is denoted by $\Gamma(\alpha, \lambda)$ or $\Gamma(x; \alpha, \lambda)$. We see that

$$\Gamma(x; \alpha, \lambda) = \begin{cases} \dfrac{\lambda^\alpha}{\Gamma(\alpha)} x^{\alpha-1}e^{-\lambda x}, & x > 0, \\ 0, & x \le 0. \end{cases} \tag{33}$$

We also record the following formula, which will prove to be useful:

$$\int_0^\infty x^{\alpha-1}e^{-\lambda x}\,dx = \frac{\Gamma(\alpha)}{\lambda^\alpha}. \tag{34}$$

The exponential densities are special cases of gamma densities. Specifically, the exponential density with parameter λ is the same as the gamma density $\Gamma(1, \lambda)$. The density given by (31) was also seen to be a gamma density with parameters $\alpha = 1/2$ and $\lambda = 1/2\sigma^2$. In other words, *if X has the normal density $n(0, \sigma^2)$, then X^2 has the gamma density $\Gamma(1/2, 1/2\sigma^2)$.* By equating (31) and (33) with $\alpha = 1/2$ and $\lambda = 1/2\sigma^2$ we obtain the useful fact that

$$\Gamma(1/2) = \sqrt{\pi}. \tag{35}$$

An important property of the gamma function is

$$\Gamma(\alpha + 1) = \alpha\Gamma(\alpha), \qquad \alpha > 0. \tag{36}$$

This formula follows from (32) by a simple application of integration by parts. To be specific

$$\begin{aligned} \Gamma(\alpha + 1) &= \int_0^\infty x^\alpha e^{-x}\,dx \\ &= -x^\alpha e^{-x}\Big|_0^\infty + \int_0^\infty \alpha x^{\alpha-1}e^{-x}\,dx \\ &= \alpha\Gamma(\alpha). \end{aligned}$$

Since $\Gamma(1) = 1$ it follows easily from (36) that if n is a positive integer,

$$\Gamma(n) = (n - 1)!. \tag{37}$$

It also follows from (35), (36) and some simplifications that if n is an odd positive integer, then

$$\Gamma\left(\frac{n}{2}\right) = \frac{\sqrt{\pi}(n - 1)!}{2^{n-1}\left(\frac{n-1}{2}\right)!}. \tag{38}$$

There are no simple formulas for the distribution function corresponding to $\Gamma(\alpha, \lambda)$ except when $\alpha = m$ is a positive integer. In this case we can integrate by parts to obtain for $x > 0$

$$\begin{aligned}\int_0^x \frac{\lambda^m y^{m-1} e^{-\lambda y}}{(m-1)!}\,dy &= \frac{-(\lambda y)^{m-1} e^{-\lambda y}}{(m-1)!}\Bigg|_0^x + \int_0^x \frac{\lambda^{m-1} y^{m-2} e^{-\lambda y}}{(m-2)!}\,dy \\ &= \int_0^x \frac{\lambda^{m-1} y^{m-2} e^{-\lambda y}}{(m-2)!}\,dy - \frac{(\lambda x)^{m-1} e^{-\lambda x}}{(m-1)!},\end{aligned}$$

provided that $m \geq 2$. If we integrate by parts $m - 1$ times in this manner and observe that

$$\int_0^x \lambda e^{-\lambda y}\,dy = 1 - e^{-\lambda x},$$

we obtain the formula

$$\int_0^x \frac{\lambda^m y^{m-1} e^{-\lambda y}}{(m-1)!}\,dy = 1 - \sum_{k=0}^{m-1} \frac{(\lambda x)^k e^{-\lambda x}}{k!}, \qquad x > 0. \tag{39}$$

This formula provides an interesting connection between a random variable X having the gamma density $\Gamma(m, \lambda)$ and a random variable Y having a Poisson distribution with parameter λx. Specifically, (39) states that

$$P(X \leq x) = P(Y \geq m). \tag{40}$$

This connection is relevant to the theory of Poisson processes, as we will see in Chapter 9.

The qualitative behavior of the gamma density, illustrated in Figure 6,

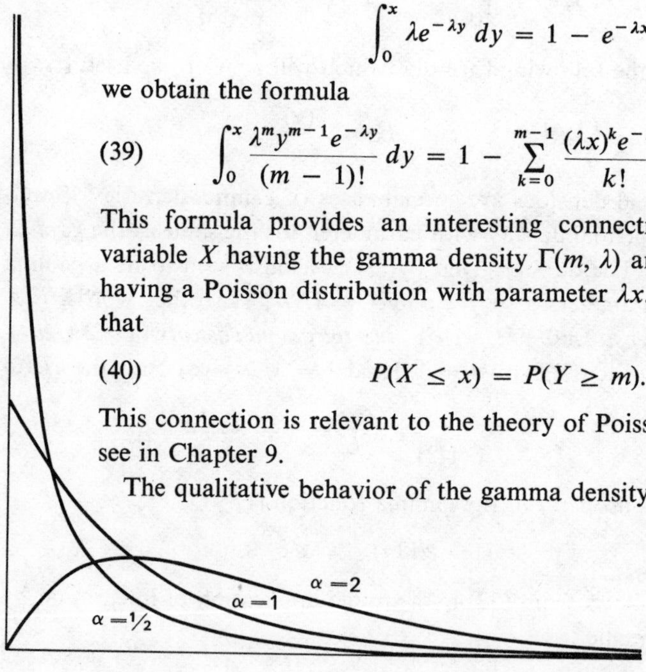

Figure 6. The Gamma Density

is easily obtained by methods of calculus. One important property of gamma densities is that if X and Y are independent random variables having respective densities $\Gamma(\alpha_1, \lambda)$ and $\Gamma(\alpha_2, \lambda)$, then $X + Y$ has the gamma density $\Gamma(\alpha_1 + \alpha_2, \lambda)$. This result will be proven in Chapter 6. This and other properties of gamma densities make them very convenient to work with. There are many applied situations when the density of a random variable X is not known. It may be known that X is a positive random variable whose density can reasonably well be approximated by a gamma density with appropriate parameters. In such cases, solving a problem involving X under the assumption that X has a gamma density will provide an approximation or at least an insight into the true but unknown situation.

5.4. Inverse distribution functions*

Important applications of the change of variable formulas of Section 5.2.1. can be obtained by letting the function φ be related to a distribution function F.

Let X be a continuous random variable having distribution function F and density function f. We will apply the change of variable formula to the function $\varphi = F$. If $y = F(x)$, then $dy/dx = F'(x) = f(x)$ and hence $dx/dy = 1/f(x)$. Thus according to (18), the random variable $Y = F(X)$ has density g where

$$g(y) = \frac{f(x)}{f(x)} = 1, \qquad 0 < y < 1,$$

and $g(y) = 0$ otherwise. In other words, the random variable $Y = F(X)$ is uniformly distributed on $(0, 1)$. This result is valid even if the function $\varphi = F$ does not satisfy all the assumptions of Theorem 1. By using a direct argument, one can show that *if X is a continuous random variable having distribution function F, then $F(X)$ is uniformly distributed on* $(0, 1)$. (If F is discontinuous at some point x_0, then $P(X = x_0) > 0$, so that $P(F(X) = F(x_0)) > 0$ and $F(X)$ could not possibly be uniformly distributed on $(0, 1)$.)

One can also proceed in the other direction. Let F be a continuous distribution function that is strictly increasing on some interval I and such that $F = 0$ to the left of I if I is bounded from below and $F = 1$ to the right of I if I is bounded from above. Then for $0 < y < 1$, by the intermediate value theorem of calculus, there is a unique value of x such that $y = F(x)$. Thus $F^{-1}(y)$, $0 < y < 1$, is well defined. Under these assumptions, *if Y is a uniformly distributed random variable on* $(0, 1)$, *then the random variable $F^{-1}(Y)$ has F as its distribution function.*

Two of the examples from Section 5.2.1 can be used to illustrate the above result. In Example 6 we obtained exponentially distributed random variables as transforms of a uniformly distributed random variable. The reader should check to see that these transformations can be obtained by the method of the above paragraph. In Example 11 we showed that if Θ is uniformly distributed on $(-\pi/2, \pi/2)$, then $\tan \Theta$ has the Cauchy distribution. Let Y be uniformly distributed on $(0, 1)$. Then $\Theta = \pi Y - \pi/2$ is uniformly distributed on $(-\pi/2, \pi/2)$, so that

$$X = \tan \Theta = \tan\left(\pi Y - \frac{\pi}{2}\right)$$

has the Cauchy distribution. This is exactly what we would get by using the result of the previous paragraph. According to Example 10, the Cauchy distribution function is given by

$$F(x) = \frac{1}{2} + \frac{1}{\pi} \arctan x, \qquad -\infty < x < \infty,$$

and the equation $y = F(x)$, or

$$y = \frac{1}{2} + \frac{1}{\pi} \arctan x,$$

has solution

$$x = F^{-1}(y) = \tan\left(\pi y - \frac{\pi}{2}\right).$$

For some purposes it is desirable to generate a random variable X having a prescribed distribution function F. One way of doing this is to first generate a uniformly distributed random variable Y and then set $X = F^{-1}(Y)$. This method is especially useful on a digital computer since there are very satisfactory methods for generating (what act like) uniformly distributed random variables on such computers. Suppose for example we want a routine for generating a random variable X having the standard normal density $n(0, 1)$. We would use a subroutine for generating a random variable Y uniformly distributed on $(0, 1)$ and a subroutine for computing the numerical function Φ^{-1}, and then compute $X = \Phi^{-1}(Y)$. To generate a random variable X having the normal density $n(\mu, \sigma^2)$ we would set $X = \mu + \sigma\Phi^{-1}(Y)$.

Inverse distribution functions are useful for other purposes. To see this let X have the normal density $n(\mu, \sigma^2)$ and recall from Section 5.3.1 that

$$P(X \le b) = \Phi\left(\frac{b - \mu}{\sigma}\right).$$

Suppose we want to choose b such that $P(X \le b) = .9$. We need to solve for b in the equation

$$\Phi\left(\frac{b - \mu}{\sigma}\right) = .9.$$

The solution is given by

$$\frac{b - \mu}{\sigma} = \Phi^{-1}(.9)$$

or

$$b = \mu + \sigma \Phi^{-1}(.9).$$

From Table I we see that $\Phi^{-1}(.9) = 1.28$. Thus $b = \mu + 1.28\sigma$ and

$$P(X \le \mu + 1.28\sigma) = .9.$$

In applied statistics the number $b = \mu + 1.28\sigma$ is called the upper decile for the $n(\mu, \sigma^2)$ distribution.

Let F be any distribution function that satisfies the requirements for $F^{-1}(y)$, $0 < y < 1$, to be well defined, as discussed above. Then $m = F^{-1}(1/2)$ is called the *median* of F, $F^{-1}(3/4)$ and $F^{-1}(1/4)$ are called the upper and lower quartiles of F, $F^{-1}(.9)$ is called the upper decile and $F^{-1}(k/100)$ is called the upper *k-percentile*. These definitions can be modified to apply to arbitrary and, in particular, discrete distribution functions.

If X has a symmetric density then X clearly has median $m = 0$. For a more interesting example, let X be exponentially distributed with parameter λ. Then its median m is given by $1 - e^{-\lambda m} = 1/2$, which has the solution $m = \lambda^{-1} \log 2$. Suppose X represents the time for a radioactive particle to decay. Then if we have a very large number of such particles we would expect that by time m one half of the particles would have decayed. In physics this time is called the *half-life* of the particle. If we observe the half-life m we can use it to compute the rate of decay λ, since $\lambda = m^{-1} \log 2$.

For a final application of inverse distribution functions, let X have the normal density $n(\mu, \sigma^2)$ and suppose we want to find $a > 0$ such that $P(\mu - a \le X \le \mu + a) = .9$. Then by (25) we have to solve for a in the equation

$$\Phi\left(\frac{a}{\sigma}\right) - \Phi\left(-\frac{a}{\sigma}\right) = .9.$$

Since $\Phi(-x) = 1 - \Phi(x)$ for all x, we have

$$2\Phi\left(\frac{a}{\sigma}\right) - 1 = .9$$

and hence $a = \sigma\Phi^{-1}(.95)$. From Table I we see that $\Phi^{-1}(.95) = 1.645$. In other words,

$$P(\mu - 1.645\sigma \le X \le \mu + 1.645\sigma) = .9.$$

By using the same technique we obtain

$$P(\mu - .675\sigma \le X \le \mu + .675\sigma) = .5$$

or equivalently,

$$P(|X - \mu| \leq .675\sigma) = .5.$$

This says that if X has the normal density $n(\mu, \sigma^2)$, then X will differ from μ by less than $.675\sigma$ with probability one-half and by more than $.675\sigma$ with probability one-half. If we think of μ as a true physical quantity and X as a measurement of μ, then $|X - \mu|$ represents the measurement error. For this reason $.675\sigma$ is known as the *probable error.*

Exercises

1 Let X be a random variable such that $P(|X - 1| = 2) = 0$. Express $P(|X - 1| \geq 2)$ in terms of the distribution function F_X.

2 Let a point be chosen randomly from the interior of a disk of radius R in the plane. Let X denote the square of the distance of the point chosen from the center of the disk. Find the distribution function of X.

3 Let a point be chosen uniformly from a solid ball in three-dimensional space of radius R. Let X denote the distance of the point chosen from the center of the ball. Find the distribution function of X.

4 Let a point be chosen uniformly over the interval $[0, a]$. Let X denote the distance of the point chosen from the origin. Find the distribution function of X.

5 Let a point be chosen uniformly from the interior of a triangle having a base of length l and height h from the base. Let X be defined as the distance from the point chosen to the base. Find the distribution function of X.

6 Consider an equilateral triangle whose sides each have length s. Let a point be chosen uniformly from one side of the triangle. Let X denote the distance of the point chosen from the opposite vertex. Find the distribution function of X.

7 Let the point (u, v) be chosen uniformly from the square $0 \leq u \leq 1$, $0 \leq v \leq 1$. Let X be the random variable that assigns to the point (u, v) the number $u + v$. Find the distribution function of X.

8 Let F be the distribution function given by Formula (3). Find a number m such that $F(m) = 1/2$.

9 Let X denote the decay time of some radioactive particle and assume that the distribution function of X is given by Formula (3). Suppose λ is such that $P(X \geq .01) = 1/2$. Find a number t such that $P(X \geq t) = .9$.

10 Let X be the random variable in Exercise 4. Find the distribution function of $Y = \text{Min}(X, a/2)$.

11 Let X be a random variable whose distribution function F is given by

$$F(x) = \begin{cases} 0, & x < 0, \\ \dfrac{x}{3}, & 0 \le x < 1, \\ \dfrac{x}{2}, & 1 \le x < 2, \\ 1, & x \ge 2. \end{cases}$$

Find:
(a) $P(1/2 \le X \le 3/2)$;
(b) $P(1/2 \le X \le 1)$;
(c) $P(1/2 \le X < 1)$;
(d) $P(1 \le X \le 3/2)$;
(e) $P(1 < X < 2)$.

12 If the distribution function of X was defined in one of the following ways, describe how properties (i)–(iv) of Section 5.1.1 would have to be modified in each case:
(a) $F(x) = P(X < x)$;
(b) $F(x) = P(X > x)$;
(c) $F(x) = P(X \ge x)$.

13 A point is chosen uniformly from $(-10, 10)$. Let X be the random variable defined so that X denotes the coordinate of the point if the point is in $[-5, 5]$, $X = -5$ if the point is in $(-10, -5)$, and $X = 5$ if the point is in $(5, 10)$. Find the distribution function of X.

14 Let X be a continuous random variable having density f given by

$$f(x) = (1/2)e^{-|x|}, \qquad -\infty < x < \infty.$$

Find $P(1 \le |X| \le 2)$.

15 Let F be the distribution function defined by

$$F(x) = \frac{1}{2} + \frac{x}{2(|x| + 1)}, \qquad -\infty < x < \infty.$$

Find a density function f for F. At what points x will $F'(x) = f(x)$?

16 Find a density function for the random variable in Exercise 3.

17 Find a density function for the random variable in Exercise 7.

18 Let X be a continuous random variable having density f. Find a formula for the density of $Y = |X|$.

19 Let X and $Y = X^2$ be positive continuous random variables having densities f and g respectively. Find f in terms of g and find g in terms of f.

20 Let X be uniformly distributed on $(0, 1)$. Find the density of $Y = X^{1/\beta}$, where $\beta \ne 0$.

21 Let X be a positive continuous random variable having density f. Find a formula for the density of $Y = 1/(X + 1)$.

22 Let X be a random variable, g a density function with respect to integration, and φ a differentiable strictly increasing function on $(-\infty, \infty)$. Suppose that

$$P(X \leq x) = \int_{-\infty}^{\varphi(x)} g(z)\, dz, \qquad -\infty < x < \infty.$$

Show that the random variable $Y = \varphi(X)$ has density g.

23 Let X be a random variable that is uniformly distributed on (a, b). Find a linear function φ such that $Y = \varphi(X)$ is uniformly distributed on $(0, 1)$.

24 Let X have an exponential density with parameter λ. Find the density of $Y = cX$, where $c > 0$.

25 Let $g(x) = x(1 - x)^2$, $0 \leq x \leq 1$, and $g(x) = 0$ elsewhere. How should g be normalized to make it a density?

26 Let X have the Cauchy density. Find the density of $Y = a + bX$, $b \neq 0$.

27 Let X denote the sine of an angle chosen at random from $(-\pi/2, \pi/2)$. Find the density and distribution function of X.

28 Let X be a continuous random variable having symmetric density f and such that X^2 has an exponential density with parameter λ. Find f.

29 Let X be a continuous random variable having distribution function F and density function f. Then f is said to be symmetric about a if $f(a + x) = f(a - x)$, $-\infty < x < \infty$. Find equivalent conditions in terms of the random variable X and in terms of the distribution function F.

30 The error function is defined by

$$erf(x) = \frac{2}{\sqrt{\pi}} \int_0^x e^{-y^2}\, dy, \qquad -\infty < x < \infty.$$

Express Φ in terms of the error function.

31 Let X have the normal density $n(0, \sigma^2)$. Find the density of $Y = |X|$.

32 Let X have the normal density $n(\mu, \sigma^2)$. Find the density of $Y = e^X$. This density is called a *lognormal density*.

33 Let X be normally distributed with parameters μ and σ^2. Find $P(|X - \mu| \leq \sigma)$.

34 Let X be normally distributed with parameters μ and σ^2. Find numbers a and b such that $a + bX$ has the standard normal distribution.

35 Let X be normally distributed with parameters $\mu = 0$ and $\sigma^2 = 4$. Let Y be the integer-valued random variable defined in terms of X by $Y = m$ if $m - 1/2 \leq X < m + 1/2$, where m is an integer such that

$-5 \le m \le 5$, $Y = -6$ if $X < -5.5$, and $Y = 6$ if $X \ge 5.5$. Find f_Y and graph this density.

36 Suppose that the weight of a person selected at random from some population is normally distributed with parameters μ and σ. Suppose also that $P(X \le 160) = 1/2$ and $P(X \le 140) = 1/4$. Find μ and σ and find $P(X \ge 200)$. Of all the people in the population weighing at least 200 pounds, what percentage will weigh over 220 pounds?

37 Let t_p be the number such that $\Phi(t_p) = p$, $0 < p < 1$. Let X have the normal density $n(\mu, \sigma^2)$. Show that for $0 < p_1 < p_2 < 1$,

$$P(\mu + t_{p_1}\sigma \le X \le \mu + t_{p_2}\sigma) = p_2 - p_1.$$

38 Suppose a very large number of identical radioactive particles have decay times which are exponentially distributed with some parameter λ. If one half of the particles decay during the first second, how long will it take for 75% of the particles to decay?

39 Let X be exponentially distributed with parameter λ. Let Y be the integer-valued random variable defined in terms of X by $Y = m$ if $m \le X < m + 1$, where m is a nonnegative integer. How is Y distributed?

40 Let T be a positive continuous random variable denoting the failure time of some system, let F denote the distribution function of T, and suppose that $F(t) < 1$ for $0 < t < \infty$. Then we can write $F(t) = 1 - e^{-G(t)}$, $t > 0$. Suppose $G'(t) = g(t)$ exists for $t > 0$.

(a) Show that T has a density f given by

$$\frac{f(t)}{1 - F(t)} = g(t), \qquad 0 < t < \infty.$$

The function g is known as the "failure rate," for heuristically,

$$P(t \le T \le t + dt \mid T > t) = \frac{f(t)\,dt}{1 - F(t)} = g(t)\,dt.$$

(b) Show that for $s > 0$ and $t > 0$,

$$P(T > t + s \mid T > t) = e^{-\int_t^{t+s} g(u)\,du}.$$

(c) Show that the system improves with age (i.e., for fixed s the expressions in (b) increase with t) if g is a decreasing function, and the system deteriorates with age if g is an increasing function.

(d) Show that

$$\int_0^\infty g(u)\,du = \infty.$$

(e) How does g behave if T is exponentially distributed?

(f) If $G(t) = \lambda t^\alpha$, $t > 0$, for which values of α does the system improve, deteriorate, and stay the same with age?

41 Let X have the gamma density $\Gamma(\alpha, \lambda)$. Find the density of $Y = cX$, where $c > 0$.

42 Show that if $\alpha > 1$, the gamma density has a maximum at $(\alpha - 1)/\lambda$.

43 Let X have the gamma density $\Gamma(\alpha, \lambda)$. Find the density of $Y = \sqrt{X}$.

44 Let Y be uniformly distributed on $(0, 1)$. Find a function φ such that $X = \varphi(Y)$ has the density f given by $f(x) = 2x$, $0 \le x \le 1$, and $f(x) = 0$ elsewhere.

45 Let Y be uniformly distributed on $(0, 1)$. Find a function φ such that $\varphi(Y)$ has the gamma density $\Gamma(1/2, 1/2)$. *Hint:* Use Example 12.

46 Find $\Phi^{-1}(t)$ for $t = .1, .2, \ldots, .9$, and use these values to graph Φ^{-1}.

47 Let X have the normal density $n(\mu, \sigma^2)$. Find the upper quartile for X.

48 Let X have the Cauchy density. Find the upper quartile for X.

49 Let X have the normal density with parameters μ and $\sigma^2 = .25$. Find a constant c such that

$$P(|X - \mu| \le c) = .9.$$

50 Let X be an integer-valued random variable having distribution function F, and let Y be uniformly distributed on $(0, 1)$. Define the integer-valued random variable Z in terms of Y by

$$Z = m \quad \text{if} \quad F(m - 1) < Y \le F(m),$$

for any integer m. Show that Z has the same density as X.

6 Jointly Distributed Random Variables

In the first three sections of this chapter we will consider a pair of continuous random variables X and Y and some of their properties. In the remaining four sections we will consider extensions from two to n random variables $X_1, X_2, \ldots, X_n$. The discussion of order statistics in Section 6.5 is optional and will not be needed later on in the book. Section 6.6 is mainly a summary of results on sampling distributions that are useful in statistics and are needed in Volume II. The material covered in Section 6.7 will be used only in proving Theorem 1 of Chapter 9 and Theorem 1 of Chapter 5 of Volume II.

6.1. Properties of bivariate distributions

Let X and Y be two random variables defined on the same probability space. Their *joint distribution function* F is defined by

$$F(x, y) = P(X \le x, Y \le y), \qquad -\infty < x, y < \infty.$$

To see that F is well defined, note that since X and Y are random variables, both $\{\omega \mid X(\omega) \le x\}$ and $\{\omega \mid Y(\omega) \le y\}$ are events. Their intersection $\{\omega \mid X(\omega) \le x \text{ and } Y(\omega) \le y\}$ is also an event, and its probability is therefore well defined.

The joint distribution function can be used to calculate the probability that the pair (X, Y) lies in a rectangle in the plane. Consider the rectangle

$$R = \{(x, y) \mid a < x \le b, c < y \le d\},$$

where $a \le b$ and $c \le d$. Then

$$(1) \qquad P((X, Y) \in R) = P(a < X \le b, c < Y \le d) = F(b, d) - F(a, d) - F(b, c) + F(a, c).$$

To verify that (1) holds observe that

$$P(a < X \le b, Y \le d) = P(X \le b, Y \le d) - P(X \le a, Y \le d) = F(b, d) - F(a, d).$$

Similarly

$$P(a < X \leq b, Y \leq c) = F(b, c) - F(a, c).$$

Thus

$$\begin{aligned} P(a < X \leq b, c < Y \leq d) &= P(a < X \leq b, Y \leq d) - P(a < X \leq b, Y \leq c) \\ &= (F(b, d) - F(a, d)) - (F(b, c) - F(a, c)) \end{aligned}$$

and (1) holds as claimed.

The one-dimensional distribution functions F_X and F_Y defined by

$$F_X(x) = P(X \leq x) \qquad \text{and} \qquad F_Y(y) = P(Y \leq y)$$

are called the *marginal distribution functions* of X and Y. They are related to the joint distribution function F by

$$F_X(x) = F(x, \infty) = \lim_{y \to \infty} F(x, y)$$

and

$$F_Y(y) = F(\infty, y) = \lim_{x \to \infty} F(x, y).$$

If there is a nonnegative function f such that

$$(2) \qquad F(x, y) = \int_{-\infty}^{x} \left(\int_{-\infty}^{y} f(u, v)\, dv \right) du, \qquad -\infty < x, y < \infty,$$

then f is called a joint density function (with respect to integration) for the distribution function F or the pair of random variables X, Y. Unless otherwise specified, throughout this chapter by density functions we shall mean density functions with respect to integration rather than discrete density functions.

If F has density f, then Equation (1) can be rewritten in terms of f, to give

$$(3) \qquad P(a < X \leq b, c < Y \leq d) = \int_a^b \left(\int_c^d f(x, y)\, dy \right) dx.$$

By using the properties of integration and the definition of a probability space, it can be shown that the relation

$$(4) \qquad P((X, Y) \in A) = \iint_A f(x, y)\, dx\, dy$$

holds for subsets A in the plane of the type considered in calculus. By letting A be the entire plane we obtain from (4) that

$$(5) \qquad \int_{-\infty}^{\infty} \int_{-\infty}^{\infty} f(x, y)\, dx\, dy = 1.$$

We also obtain from (4) that

$$F_X(x) = P(X \leq x) = \int_{-\infty}^{x} \left(\int_{-\infty}^{\infty} f(u, y)\, dy \right) du$$

and hence F_X has marginal density f_X given by

$$f_X(x) = \int_{-\infty}^{\infty} f(x, y)\, dy$$

which satisfies

$$F_X(x) = \int_{-\infty}^{x} f_X(u)\, du.$$

Similarly F_Y has marginal density f_Y given by

$$f_Y(y) = \int_{-\infty}^{\infty} f(x, y)\, dx.$$

As in the one-dimensional case, f is not uniquely defined by (2). We can change f at a finite number of points or even over a finite number of smooth curves in the plane without affecting integrals of f over sets in the plane. Again as in the one-dimensional case, F determines f at the continuity points of f. This fact can be obtained from (3).

By differentiating (2) and applying the rules of calculus we obtain

$$\frac{\partial}{\partial y} F(x, y) = \int_{-\infty}^{x} \left(\frac{\partial}{\partial y} \int_{-\infty}^{y} f(u, v)\, dv \right) du$$

$$= \int_{-\infty}^{x} f(u, y)\, du$$

and

(6) $$\frac{\partial^2}{\partial x\, \partial y} F(x, y) = f(x, y).$$

Under some further mild conditions we can justify these operations and show that (6) holds at the continuity points of f. In specific cases instead of checking that the steps leading to (6) are valid, it is usually simpler to show that the function f obtained from (6) satisfies (2).

Example 1. Let us illustrate the above definitions and formulas by reconsidering Example 1 of Chapter 5. We recall that in that example, we chose a point uniformly from a disk of radius R. Let points in the plane be determined by their Cartesian coordinates (x, y). Then the disk can be written as

$$\{(x, y) \mid x^2 + y^2 \leq R^2\}.$$

Let X and Y be random variables denoting the random coordinates of the point chosen. Corresponding to the assumption of uniformity, we suppose that X and Y have a joint density f given by

(7) $$f(x, y) = \begin{cases} \dfrac{1}{\pi R^2}, & x^2 + y^2 \le R^2, \\ 0, & \text{elsewhere.} \end{cases}$$

Then for any subset A of the disk (say of the type considered in calculus),

$$P((X, Y) \in A) = \iint_A f(x, y)\, dx\, dy$$

$$= \frac{\text{area of } A}{\pi R^2},$$

which agrees with our assumption of uniformity. The marginal density f_X is given by

$$f_X(x) = \int_{-\infty}^{\infty} f(x, y)\, dy = \int_{-\sqrt{R^2 - x^2}}^{\sqrt{R^2 - x^2}} \frac{1}{\pi R^2}\, dy = \frac{2\sqrt{R^2 - x^2}}{\pi R^2}$$

for $-R < x < R$ and $f_X(x) = 0$ elsewhere. The marginal density $f_Y(y)$ is given by the same formula with x replaced by y.

The variables X and Y are called *independent random variables* if whenever $a \le b$ and $c \le d$, then

(8) $$P(a < X \le b, c < Y \le d) = P(a < X \le b)P(c < Y \le d).$$

By letting $a = c = -\infty$, $b = x$, and $d = y$, it follows that if X and Y are independent, then

(9) $$F(x, y) = F_X(x)F_Y(y), \qquad -\infty < x, y < \infty.$$

Conversely (9) implies that X and Y are independent. For if (9) holds, then by (1) the left side of (8) is

$$\begin{aligned} F(b, d) - F(a, d) - F(b, c) + F(a, c) \\ &= F_X(b)F_Y(d) - F_X(a)F_Y(d) - F_X(b)F_Y(c) + F_X(a)F_Y(c) \\ &= (F_X(b) - F_X(a))(F_Y(d) - F_Y(c)) \\ &= P(a < X \le b)P(c < Y \le d). \end{aligned}$$

More generally, it can be shown that if X and Y are independent and A and B are unions of a finite or countably infinite number of intervals, then

$$P(X \in A, Y \in B) = P(X \in A)P(Y \in B)$$

or, in other words, the events

$$\{\omega \mid X(\omega) \in A\} \quad \text{and} \quad \{\omega \mid X(\omega) \in B\}$$

are independent events.

Let X and Y be random variables having marginal densities f_X and f_Y. Then X and Y are independent if and only if the function f defined by

$$f(x, y) = f_X(x) f_Y(y), \qquad -\infty < x, y < \infty,$$

is a joint density for X and Y. This follows from the definition of independence and the formula

$$F_X(x)F_Y(y) = \int_{-\infty}^{x} \left(\int_{-\infty}^{y} f_X(u)f_Y(v)\, dv \right) du.$$

As an illustration of dependent random variables, let X and Y be as in Example 1. Then for $-R < x < R$ and $-R < y < R$,

$$f_X(x)f_Y(y) = \frac{4\sqrt{R^2 - x^2}\sqrt{R^2 - y^2}}{\pi^2 R^4}, \tag{10}$$

which does not agree with the joint density of these variables at $x = 0$, $y = 0$. Since $(0, 0)$ is a continuity point of the functions defined by (7) and (10), it follows that X and Y are dependent random variables. This agrees with our intuitive notion of dependence since when X is close to R, Y must be close to zero, so information about X gives us information about Y.

Density functions can also be defined directly, as we have seen in other contexts. A *two-dimensional* (*or bivariate*) *density function* f is a nonnegative function on R^2 such that

$$\int_{-\infty}^{\infty} \int_{-\infty}^{\infty} f(x, y)\, dx\, dy = 1.$$

Corresponding to any bivariate density function f, there is a probability space and a pair of random variables X and Y defined on that space and having joint density f.

The easiest way to construct two-dimensional density functions is to start with two one-dimensional densities f_1 and f_2 and define the function f by

$$f(x, y) = f_1(x)f_2(y), \qquad -\infty < x, y < \infty. \tag{11}$$

Then f is a two-dimensional density function since it is clearly nonnegative and

$$\int_{-\infty}^{\infty} \int_{-\infty}^{\infty} f(x, y)\, dx\, dy = \int_{-\infty}^{\infty} f_1(x)\, dx \int_{-\infty}^{\infty} f_2(y)\, dy = 1.$$

If random variables X and Y have this f as their joint density, then X and Y are independent and have marginal densities $f_X = f_1$ and $f_Y = f_2$.

As an illustration of (11), let f_1 and f_2 both be the standard normal density $n(0, 1)$. Then f is given by

$$f(x, y) = \frac{1}{\sqrt{2\pi}} e^{-x^2/2} \frac{1}{\sqrt{2\pi}} e^{-y^2/2}$$

or

$$(12) \qquad f(x, y) = \frac{1}{2\pi} e^{-(x^2+y^2)/2}, \qquad -\infty < x, y < \infty.$$

The density given by (12) is called the *standard bivariate normal density*. In our next example we will modify the right side of (12) slightly to obtain a joint density function that corresponds to the case where the two random variables having normal marginal densities are dependent.

Example 2. Let X and Y have the joint density function f given by

$$f(x, y) = ce^{-(x^2-xy+y^2)/2}, \qquad -\infty < x, y < \infty,$$

where c is a positive constant that will be determined in the course of our discussion. We first "complete the square" in the terms involving y and rewrite f as

$$f(x, y) = ce^{-[(y-x/2)^2+3x^2/4]/2}, \qquad -\infty < x, y < \infty,$$

and then note that

$$f_X(x) = \int_{-\infty}^{\infty} f(x, y)\, dy = ce^{-3x^2/8} \int_{-\infty}^{\infty} e^{-(y-x/2)^2/2}\, dy.$$

Making the change of variable $u = y - x/2$, we see that

$$\int_{-\infty}^{\infty} e^{-(y-x/2)^2/2}\, dy = \int_{-\infty}^{\infty} e^{-u^2/2}\, du = \sqrt{2\pi}.$$

Consequently

$$f_X(x) = c\sqrt{2\pi}e^{-3x^2/8}.$$

It is now clear that f_X is the normal density $n(0, \sigma^2)$ with $\sigma^2 = 4/3$ and hence

$$c\sqrt{2\pi} = \frac{1}{\sigma\sqrt{2\pi}} = \frac{\sqrt{3}}{2\sqrt{2\pi}}$$

or $c = \sqrt{3}/4\pi$. Consequently

$$(13) \qquad f(x, y) = \frac{\sqrt{3}}{4\pi} e^{-(x^2-xy+y^2)/2}, \qquad -\infty < x, y < \infty.$$

The above calculations now show that f_X is the normal density $n(0, 4/3)$. In a similar fashion, we can show that f_Y is also $n(0, 4/3)$. Since $f(x, y) \neq f_X(x)f_Y(y)$, it is clear that X and Y are dependent.

6.2. Distribution of sums and quotients

Let X and Y be random variables having joint density f. In many contexts we have a random variable Z defined in terms of X and Y and we wish to calculate the density of Z. Let Z be given by $Z = \varphi(X, Y)$, where φ is a real-valued function whose domain contains the range of X and Y. For fixed z the event $\{Z \le z\}$ is equivalent to the event $\{(X, Y) \in A_z\}$, where A_z is the subset of R^2 defined by

$$A_z = \{(x, y) \mid \varphi(x, y) \le z\}.$$

Thus

$$\begin{aligned} F_Z(z) &= P(Z \le z) \\ &= P((X, Y) \in A_z) \\ &= \iint_{A_z} f(x, y)\, dx\, dy. \end{aligned}$$

If we can find a nonnegative function g such that

$$\iint_{A_z} f(x, y)\, dx\, dy = \int_{-\infty}^{z} g(v)\, dv, \qquad -\infty < z < \infty,$$

then g is necessarily a density of Z. We will use this method to calculate densities of $X + Y$ and Y/X.

6.2.1. Distribution of sums.

Set $Z = X + Y$. Then

$$A_z = \{(x, y) \mid x + y \le z\}$$

is just the half-plane to the lower left of the line $x + y = z$ as shown in Figure 1. Thus

$$F_Z(z) = \iint_{A_z} f(x, y)\, dx\, dy = \int_{-\infty}^{\infty} \left(\int_{-\infty}^{z-x} f(x, y)\, dy \right) dx.$$

Make the change of variable $y = v - x$ in the inner integral. Then

$$\begin{aligned} F_Z(z) &= \int_{-\infty}^{\infty} \left(\int_{-\infty}^{z} f(x, v - x)\, dv \right) dx \\ &= \int_{-\infty}^{z} \left(\int_{-\infty}^{\infty} f(x, v - x)\, dx \right) dv, \end{aligned}$$

where we have interchanged the order of integration. Thus the density of $Z = X + Y$ is given by

$$(14) \qquad f_{X+Y}(z) = \int_{-\infty}^{\infty} f(x, z - x)\, dx, \qquad -\infty < z < \infty.$$

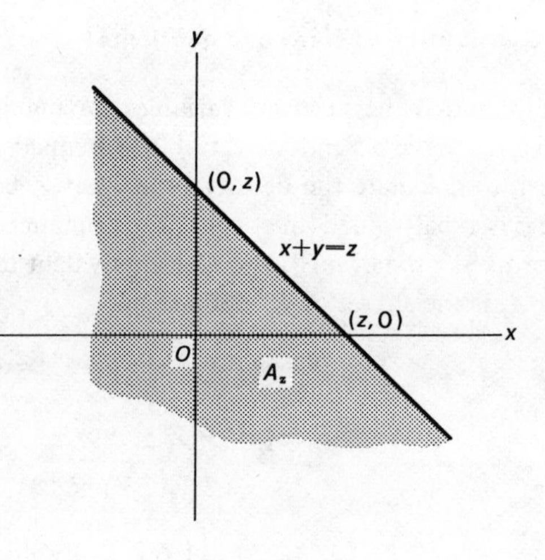

Figure 1

In the main applications of (14), X and Y are independent and (14) can be rewritten as

$$(15) \qquad f_{X+Y}(z) = \int_{-\infty}^{\infty} f_X(x) f_Y(z-x)\, dx, \qquad -\infty < z < \infty.$$

If X and Y are nonnegative independent random variables, then $f_{X+Y}(z) = 0$ for $z \leq 0$ and

$$(16) \qquad f_{X+Y}(z) = \int_0^z f_X(x) f_Y(z-x)\, dx, \qquad 0 < z < \infty.$$

The right side of (15) suggests a method of obtaining densities. Given two one-dimensional densities f and g, the function h defined by

$$h(z) = \int_{-\infty}^{\infty} f(x) g(z-x)\, dx, \qquad -\infty < z < \infty,$$

is a one-dimensional density function, which is called the *convolution* of f and g. Thus the density of the sum of two independent random variables is the convolution of the individual densities.

Example 3. Let X and Y be independent random variables each having an exponential distribution with parameter λ. Find the distribution of $X + Y$.

The density of X is given by $f_X(x) = \lambda e^{-\lambda x}$ for $x \geq 0$ and $f_X(x) = 0$ for $x < 0$. The density of Y is the same. Thus $f_{X+Y}(z) = 0$ for $z \leq 0$ and, by (16), for $z > 0$

$$\begin{aligned} f_{X+Y}(z) &= \int_0^z \lambda e^{-\lambda x} \lambda e^{-\lambda(z-x)}\, dx \\ &= \lambda^2 e^{-\lambda z} \int_0^z dx = \lambda^2 z e^{-\lambda z}. \end{aligned}$$

We see that $X + Y$ has the gamma density $\Gamma(2, \lambda)$.

Example 4. Let X and Y be independent and uniformly distributed over $(0, 1)$. Find the density of $X + Y$.

The density of X is given by $f_X(x) = 1$ for $0 < x < 1$ and $f_X(x) = 0$ elsewhere. The density of Y is the same. Thus $f_{X+Y}(z) = 0$ for $z \leq 0$. For $z > 0$ we apply (16). The integrand $f_X(x)f_Y(z - x)$ takes on only the values 0 and 1. It takes on the value 1 if x and z are such that $0 \leq x \leq 1$ and $0 \leq z - x \leq 1$. If $0 \leq z \leq 1$, the integrand has value 1 on the set $0 \leq x \leq z$ and zero otherwise. Therefore we obtain from (16) that

$$f_{X+Y}(z) = z, \qquad 0 \leq z \leq 1.$$

If $1 < z \leq 2$ the integrand has value 1 on the set $z - 1 \leq x \leq 1$ and zero otherwise. Thus by (16)

$$f_{X+Y}(z) = 2 - z, \qquad 1 < z \leq 2.$$

If $2 < z < \infty$ the integrand in (16) is identically zero and hence

$$f_{X+Y}(z) = 0, \qquad 2 < z < \infty.$$

In summary

$$f_{X+Y}(z) = \begin{cases} z, & 0 \leq z \leq 1, \\ 2 - z, & 1 < z \leq 2, \\ 0, & \text{elsewhere.} \end{cases}$$

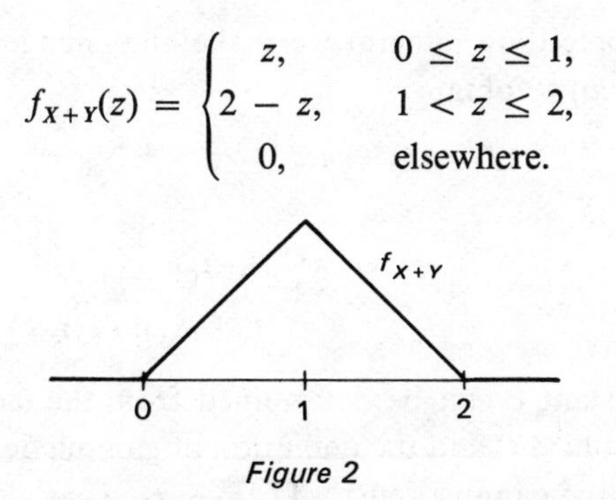

Figure 2

The graph of f is given in Figure 2. One can also find the density of $X + Y$ by computing the area of the set

$$A_z = \{(x, y) \mid 0 \leq x \leq 1, 0 \leq y \leq 1 \quad \text{and} \quad x + y \leq z\}$$

(see Figure 3) and differentiating the answer with respect to z.

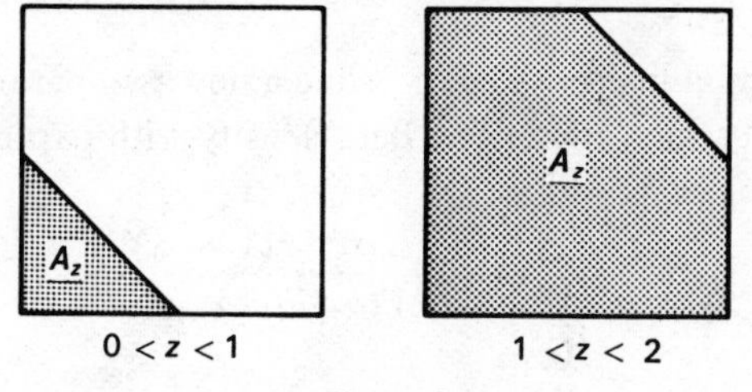

Figure 3

Example 3 has an important generalization, which may be stated as follows.

Theorem 1 *Let X and Y be independent random variables such that X has the gamma density $\Gamma(\alpha_1, \lambda)$ and Y has the gamma density $\Gamma(\alpha_2, \lambda)$. Then $X + Y$ has the gamma density*

$$\Gamma(\alpha_1 + \alpha_2, \lambda).$$

Proof. We note that X and Y are positive random variables and that

$$f_X(x) = \frac{\lambda^{\alpha_1} x^{\alpha_1 - 1} e^{-\lambda x}}{\Gamma(\alpha_1)}, \qquad x > 0,$$

and

$$f_Y(y) = \frac{\lambda^{\alpha_2} y^{\alpha_2 - 1} e^{-\lambda y}}{\Gamma(\alpha_2)}, \qquad y > 0.$$

Thus $f_{X+Y}(z) = 0$ for $z \leq 0$ and, by (16), for $z > 0$

$$f_{X+Y}(z) = \frac{\lambda^{\alpha_1+\alpha_2} e^{-\lambda z}}{\Gamma(\alpha_1)\Gamma(\alpha_2)} \int_0^z x^{\alpha_1 - 1}(z - x)^{\alpha_2 - 1}\, dx.$$

In the preceding integral we make the change of variable $x = zu$ (with $dx = z\, du$) to obtain

$$f_{X+Y}(z) = c\lambda^{\alpha_1+\alpha_2} z^{\alpha_1+\alpha_2-1} e^{-\lambda z}, \qquad z > 0, \tag{17}$$

where

$$c = \frac{\int_0^1 u^{\alpha_1 - 1}(1 - u)^{\alpha_2 - 1}\, du}{\Gamma(\alpha_1)\Gamma(\alpha_2)}. \tag{18}$$

The constant c can be determined from the fact that f_{X+Y} integrates out to 1. From (17) and the definition of gamma densities, it is clear that f_{X+Y} must be the gamma density $\Gamma(\alpha_1 + \alpha_2, \lambda)$ as claimed. ∎

From (17) and the definition of the gamma density we also see that $c = 1/\Gamma(\alpha_1 + \alpha_2)$. This together with (18) allows us to evaluate the definite integral appearing in (18) in terms of the gamma function:

$$\int_0^1 u^{\alpha_1 - 1}(1 - u)^{\alpha_2 - 1}\, du = \frac{\Gamma(\alpha_1)\Gamma(\alpha_2)}{\Gamma(\alpha_1 + \alpha_2)}. \tag{19}$$

This formula permits us to define a new two parameter family of densities called *Beta* densities. The Beta density with parameters α_1 and α_2 is given by

$$f(x) = \begin{cases} \dfrac{\Gamma(\alpha_1 + \alpha_2) x^{\alpha_1 - 1}(1 - x)^{\alpha_2 - 1}}{\Gamma(\alpha_1)\Gamma(\alpha_2)}, & 0 < x < 1, \\ 0, & \text{elsewhere}. \end{cases} \tag{20}$$

The reason for this terminology is that the function of α_1 and α_2 defined by

$$B(\alpha_1, \alpha_2) = \frac{\Gamma(\alpha_1)\Gamma(\alpha_2)}{\Gamma(\alpha_1 + \alpha_2)}, \qquad 0 < \alpha_1, \alpha_2 < \infty,$$

is called the Beta function.

Our final application of the convolution formula is to normally distributed random variables.

Theorem 2 *Let X and Y be independent random variables having the respective normal densities $n(\mu_1, \sigma_1^2)$ and $n(\mu_2, \sigma_2^2)$. Then $X + Y$ has the normal density*

$$n(\mu_1 + \mu_2, \sigma_1^2 + \sigma_2^2).$$

Proof. We assume first that $\mu_1 = \mu_2 = 0$. Then

$$f_X(x) = \frac{1}{\sigma_1\sqrt{2\pi}} e^{-x^2/2\sigma_1^2}, \qquad -\infty < x < \infty,$$

and

$$f_Y(y) = \frac{1}{\sigma_2\sqrt{2\pi}} e^{-y^2/2\sigma_2^2}, \qquad -\infty < y < \infty.$$

Thus by (15)

$$f_{X+Y}(z) = \frac{1}{2\pi\sigma_1\sigma_2} \int_{-\infty}^{\infty} \exp\left[-\frac{1}{2}\left(\frac{x^2}{\sigma_1^2} + \frac{(z-x)^2}{\sigma_2^2}\right)\right] dx.$$

Unfortunately an evaluation of this integral requires some messy computations (which are not important enough to master). One way of proceeding is to first make the change of variable

$$u = \frac{\sqrt{\sigma_1^2 + \sigma_2^2}}{\sigma_1\sigma_2} x.$$

After some simple algebra we find that

$$f_{X+Y}(z) = \frac{1}{2\pi\sqrt{\sigma_1^2 + \sigma_2^2}} \int_{-\infty}^{\infty} \exp\left[-\frac{1}{2}\left(u^2 - \frac{2uz\sigma_1}{\sigma_2\sqrt{\sigma_1^2 + \sigma_2^2}} + \frac{z^2}{\sigma_2^2}\right)\right] du.$$

We next complete the square in u and observe that

$$u^2 - \frac{2uz\sigma_1}{\sigma_2\sqrt{\sigma_1^2 + \sigma_2^2}} + \frac{z^2}{\sigma_2^2} = \left(u - \frac{z\sigma_1}{\sigma_2\sqrt{\sigma_1^2 + \sigma_2^2}}\right)^2 + \frac{z^2}{\sigma_1^2 + \sigma_2^2}.$$

Then by making a second change of variable

$$v = u - \frac{z\sigma_1}{\sigma_2\sqrt{\sigma_1^2 + \sigma_2^2}},$$

we see that

$$f_{X+Y}(z) = \frac{e^{-z^2/2(\sigma_1^2+\sigma_2^2)}}{\sqrt{2\pi}\sqrt{\sigma_1^2 + \sigma_2^2}} \int_{-\infty}^{\infty} \frac{e^{-v^2/2}}{\sqrt{2\pi}}\, dv$$

$$= \frac{e^{-z^2/2(\sigma_1^2+\sigma_2^2)}}{\sqrt{2\pi}\sqrt{\sigma_1^2 + \sigma_2^2}},$$

which is just the normal density $n(0, \sigma_1^2 + \sigma_2^2)$.

In general, $X - \mu_1$ and $Y - \mu_2$ are independent and have the respective normal densities $n(0, \sigma_1^2)$ and $n(0, \sigma_2^2)$. Thus by the above special case, $(X - \mu_1) + (Y - \mu_2) = X + Y - (\mu_1 + \mu_2)$ has the normal density $n(0, \sigma_1^2 + \sigma_2^2)$, and hence $X + Y$ has the normal density

$$n(\mu_1 + \mu_2, \sigma_1^2 + \sigma_2^2)$$

as claimed. ∎

The preceding proof is elementary but messy. A less computational proof involving more advanced techniques will be given in Section 8.3. Another proof is indicated in Exercise 36 at the end of this chapter.

Example 5. Let X and Y be independent random variables each having the normal density $n(0, \sigma^2)$. Find the density of $X + Y$ and $X^2 + Y^2$.

From Theorem 2 we see immediately that $X + Y$ has the normal density $n(0, 2\sigma^2)$. By Example 12 of Chapter 5, X^2 and Y^2 each have the gamma density $\Gamma(1/2, 1/2\sigma^2)$. It is easily seen that X^2 and Y^2 are independent. Thus by Theorem 1, $X^2 + Y^2$ has the gamma density $\Gamma(1, 1/2\sigma^2)$, which is the same as the exponential density with parameter $1/2\sigma^2$.

6.2.2. Distribution of quotients*. As before, let X and Y denote random variables having joint density f. We will now derive a formula for the density of the random variable $Z = Y/X$. The set

$$A_z = \{(x, y) \mid y/x \le z\}$$

is shown in Figure 4. If $x < 0$, then $y/x \le z$ if and only if $y \ge xz$. Thus

$$A_z = \{(x, y) \mid x < 0 \text{ and } y \ge xz\} \cup \{(x, y) \mid x > 0 \text{ and } y \le xz\}.$$

Consequently

$$F_{Y/X}(z) = \iint_{A_z} f(x, y)\, dx\, dy$$

$$= \int_{-\infty}^{0} \left(\int_{xz}^{\infty} f(x, y)\, dy\right) dx + \int_{0}^{\infty} \left(\int_{-\infty}^{xz} f(x, y)\, dy\right) dx.$$

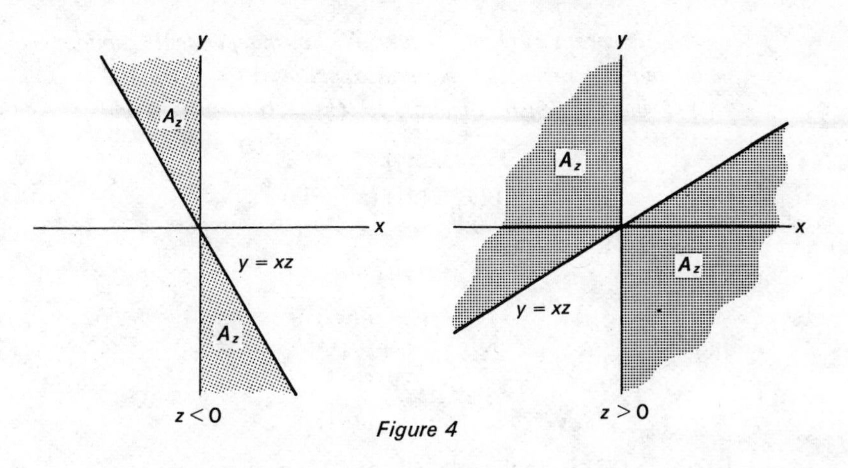

Figure 4

In the inner integrals, we make the change of variable $y = xv$ (with $dy = x\,dv$) to obtain

$$\begin{aligned} F_{Y/X}(z) &= \int_{-\infty}^{0} \left(\int_{z}^{-\infty} xf(x, xv)\, dv \right) dx \\ &\quad + \int_{0}^{\infty} \left(\int_{-\infty}^{z} xf(x, xv)\, dv \right) dx \\ &= \int_{-\infty}^{0} \left(\int_{-\infty}^{z} (-x)f(x, xv)\, dv \right) dx \\ &\quad + \int_{0}^{\infty} \left(\int_{-\infty}^{z} xf(x, xv)\, dv \right) dx \\ &= \int_{-\infty}^{\infty} \left(\int_{-\infty}^{z} |x| f(x, xv)\, dv \right) dx. \end{aligned}$$

By interchanging the order of integration we see that

$$(21) \quad F_{Y/X}(z) = \int_{-\infty}^{z} \left(\int_{-\infty}^{\infty} |x| f(x, xv)\, dx \right) dv, \qquad -\infty < z < \infty.$$

It follows from (21) that Y/X has the density $f_{Y/X}$ given by

$$(22) \qquad f_{Y/X}(z) = \int_{-\infty}^{\infty} |x| f(x, xz)\, dx, \qquad -\infty < z < \infty.$$

In the special case when X and Y are independent positive random variables, (22) reduces to $f_{Y/X}(z) = 0$ for $z \le 0$ and

$$(23) \qquad f_{Y/X}(z) = \int_{0}^{\infty} xf_X(x)f_Y(xz)\, dx, \qquad 0 < z < \infty.$$

Our next theorem is a direct application of (23).

Theorem 3 *Let X and Y be independent random variables having the respective gamma densities $\Gamma(\alpha_1, \lambda)$ and $\Gamma(\alpha_2, \lambda)$. Then Y/X has the density given by $f_{Y/X}(z) = 0$ for $z \le 0$ and*

$$(24) \qquad f_{Y/X}(z) = \frac{\Gamma(\alpha_1 + \alpha_2)}{\Gamma(\alpha_1)\Gamma(\alpha_2)} \frac{z^{\alpha_2 - 1}}{(z + 1)^{\alpha_1 + \alpha_2}}, \qquad 0 < z < \infty.$$

Proof. Recall that

$$f_X(x) = \frac{\lambda^{\alpha_1} x^{\alpha_1 - 1} e^{-\lambda x}}{\Gamma(\alpha_1)}, \qquad x > 0,$$

and

$$f_Y(y) = \frac{\lambda^{\alpha_2} y^{\alpha_2 - 1} e^{-\lambda y}}{\Gamma(\alpha_2)}, \qquad y > 0.$$

Formula (23) is applicable, so for $0 < z < \infty$,

$$\begin{aligned} f_{Y/X}(z) &= \frac{\lambda^{\alpha_1 + \alpha_2}}{\Gamma(\alpha_1)\Gamma(\alpha_2)} \int_0^\infty x x^{\alpha_1 - 1} e^{-\lambda x} (xz)^{\alpha_2 - 1} e^{-\lambda x z}\, dx \\ &= \frac{\lambda^{\alpha_1 + \alpha_2} z^{\alpha_2 - 1}}{\Gamma(\alpha_1)\Gamma(\alpha_2)} \int_0^\infty x^{\alpha_1 + \alpha_2 - 1} e^{-x\lambda(z+1)}\, dx. \end{aligned}$$

By Equation (34) of Chapter 5

$$\int_0^\infty x^{\alpha_1 + \alpha_2 - 1} e^{-x\lambda(z+1)}\, dx = \frac{\Gamma(\alpha_1 + \alpha_2)}{(\lambda(z + 1))^{\alpha_1 + \alpha_2}}.$$

Consequently (24) holds as claimed. ∎

Since (24) defines a density function we see that for $\alpha_1, \alpha_2 > 0$

$$\int_0^\infty z^{\alpha_2 - 1}(z + 1)^{-(\alpha_1 + \alpha_2)}\, dz = \frac{\Gamma(\alpha_1)\Gamma(\alpha_2)}{\Gamma(\alpha_1 + \alpha_2)}.$$

Example 6. Let X and Y be independent random variables each having the normal density $n(0, \sigma^2)$. Find the density of Y^2/X^2.

The random variables are the same as those of Example 5. Thus again X^2 and Y^2 are independent, and each has the gamma density $\Gamma(1/2, 1/2\sigma^2)$. Theorem 3 is now applicable and Y^2/X^2 has the density f_{Y^2/X^2} given by $f_{Y^2/X^2}(z) = 0$ for $z \le 0$ and

$$\begin{aligned} f_{Y^2/X^2}(z) &= \frac{\Gamma(1)}{\Gamma(1/2)\Gamma(1/2)} \frac{z^{-1/2}}{(z + 1)} \\ &= \frac{1}{\pi(z + 1)\sqrt{z}}, \qquad 0 < z < \infty. \end{aligned}$$

(Here we recall from Equation (35) of Chapter 5 that $\Gamma(1/2) = \sqrt{\pi}$.) We leave it to the reader to show as an exercise that under the same conditions both Y/X and $Y/|X|$ have the Cauchy density.

6.3. Conditional densities

In order to motivate the definition of conditional densities of continuous random variables, we will first discuss discrete random variables. Let X and Y be discrete random variables having joint density f. If x is a possible value of X, then

$$P(Y = y \mid X = x) = \frac{P(X = x, Y = y)}{P(X = x)} = \frac{f(x, y)}{f_X(x)}.$$

The function $f_{Y|X}$ defined by

$$f_{Y|X}(y \mid x) = \begin{cases} \dfrac{f(x, y)}{f_X(x)}, & f_X(x) \neq 0, \\ 0, & f_X(x) = 0, \end{cases} \tag{25}$$

is called the conditional density of Y given X. For any possible value x of X,

$$\sum_y f_{Y|X}(y \mid x) = \frac{\sum_y f(x, y)}{f_X(x)} = \frac{f_X(x)}{f_X(x)} = 1,$$

so that for any such x, $f_{Y|X}(y \mid x)$ defines a discrete density function of y known as the conditional density of Y given $X = x$. In the discrete case conditional densities involve no really new concepts.

If X is a continuous random variable, however, then $P(X = x) = 0$ for all x so that $P(Y = y \mid X = x)$ is always undefined. In this case any definition of conditional densities necessarily involves a new concept. The simplest way to define conditional densities of continuous random variables is by analogy with Formula (25) in the discrete case.

Definition 1 *Let X and Y be continuous random variables having joint density f. The conditional density $f_{Y|X}$ is defined by*

$$f_{Y|X}(y \mid x) = \begin{cases} \dfrac{f(x, y)}{f_X(x)}, & 0 < f_X(x) < \infty, \\ 0, & \textit{elsewhere}. \end{cases} \tag{26}$$

It follows immediately from this definition that, as a function of y, $f_{Y|X}(y \mid x)$ is a density whenever $0 < f_X(x) < \infty$ (again called the conditional density of Y given $X = x$). Conditional densities can be used to define conditional probabilities. Thus we define

$$P(a \le Y \le b \mid X = x) = \int_a^b f_{Y|X}(y \mid x)\, dy, \qquad a \le b. \tag{27}$$

Alternatively, we could attempt to define the conditional probability appearing in (27) by means of the following limit:

$$P(a \le Y \le b \mid X = x) = \lim_{h \downarrow 0} P(a \le Y \le b \mid x - h \le X \le x + h). \tag{28}$$

The right-hand side of (28) can be rewritten in terms of f as

$$\lim_{h \downarrow 0} \frac{\int_{x-h}^{x+h} \left(\int_a^b f(u, y)\, dy\right) du}{\int_{x-h}^{x+h} \left(\int_{-\infty}^{\infty} f(u, y)\, dy\right) du} = \lim_{h \downarrow 0} \frac{(1/2h) \int_{x-h}^{x+h} \left(\int_a^b f(u, y)\, dy\right) du}{(1/2h) \int_{x-h}^{x+h} f_X(u)\, du}.$$

If

$$\int_a^b f(u, y)\, dy$$

is continuous in u at $u = x$, the numerator of the last limit converges to

$$\int_a^b f(x, y)\, dy$$

as $h \downarrow 0$. If f_X is continuous at x the denominator converges to $f_X(x)$ as $h \downarrow 0$. Under the additional condition that $f_X(x) \neq 0$, we are led from (28) to

$$P(a \leq Y \leq b \mid X = x) = \frac{\int_a^b f(x, y)\, dy}{f_X(x)},$$

which agrees with (27). In summary, we have defined conditional densities and conditional probabilities in the continuous case by analogy with the discrete case. We have also noted that, under further restrictions, a limiting process would yield the same definition of conditional probabilities. It turns out that such limiting processes are difficult to work with and will not be used further.

It follows immediately from the definition of conditional density functions that

$$(29) \qquad f(x, y) = f_X(x) f_{Y|X}(y \mid x), \qquad -\infty < x, y < \infty.$$

If X and Y are independent and

$$(30) \qquad f(x, y) = f_X(x) f_Y(y), \qquad -\infty < x, y < \infty,$$

then

$$(31) \quad f_{Y|X}(y \mid x) = f_Y(y), \qquad 0 < f_X(x) < \infty \text{ and } -\infty < y < \infty.$$

Conversely if (31) holds, then it follows from (29) that (30) holds and X and Y are independent. Thus (31) is a necessary and sufficient condition for two random variables X and Y having a joint density to be independent.

Example 7. Let X and Y have the bivariate density f given by Formula (13), namely

$$f(x, y) = \frac{\sqrt{3}}{4\pi} e^{-(x^2 - xy + y^2)/2}, \qquad -\infty < x, y < \infty.$$

Then as we saw in Example 2, X has the normal density $n(0, 4/3)$. Thus for $-\infty < x, y < \infty$

$$f_{Y|X}(y \mid x) = \frac{\dfrac{\sqrt{3}}{4\pi} e^{-(x^2 - xy + y^2)/2}}{\dfrac{\sqrt{3}}{2\sqrt{2\pi}} e^{-3x^2/8}}$$

$$= \frac{1}{\sqrt{2\pi}} e^{-(y - x/2)^2/2}.$$

In other words, the conditional density of Y given $X = x$ is the normal density $n(x/2, 1)$.

We have been starting with joint densities and using them to construct marginal densities and conditional densities. In some situations we may reverse this by starting with marginal densities and conditional densities and using them to construct joint densities.

Example 8. Let X be a uniformly distributed random variable over $(0, 1)$, and let Y be a uniformly distributed random variable over $(0, X)$. Find the joint density of X and Y and the marginal density of Y.

From the statement of the problem, we see that the marginal density of X is given by

$$f_X(x) = \begin{cases} 1 & \text{for} \quad 0 < x < 1, \\ 0, & \quad \text{elsewhere.} \end{cases}$$

The density of Y given $X = x$ is uniform on $(0, x)$, so that

$$f_{Y|X}(y \mid x) = \begin{cases} 1/x & \text{for} \quad 0 < y < x < 1, \\ 0, & \quad \text{elsewhere.} \end{cases}$$

Thus the joint density of X and Y is given by

$$f(x, y) = \begin{cases} 1/x & \text{for} \quad 0 < y < x < 1, \\ 0, & \quad \text{elsewhere.} \end{cases}$$

The marginal density of Y is

$$f_Y(y) = \int_{-\infty}^{\infty} f(x, y)\, dx = \int_y^1 \frac{1}{x}\, dx = -\log y, \qquad 0 < y < 1,$$

and $f_Y(y) = 0$ elsewhere.

6.3.1. Bayes' rule. Of course, we can reverse the roles of X and Y and define the conditional density of X given $Y = y$ by means of the formula

$$(32) \qquad f_{X|Y}(x \mid y) = \frac{f(x, y)}{f_Y(y)}, \qquad 0 < f_Y(y) < \infty.$$

Since

$$f(x, y) = f_X(x) f_{Y|X}(y \mid x)$$

and

$$f_Y(y) = \int_{-\infty}^{\infty} f(x, y)\, dx = \int_{-\infty}^{\infty} f_X(x) f_{Y|X}(y \mid x)\, dx,$$

we can rewrite (32) as

$$f_{X|Y}(x \mid y) = \frac{f_X(x) f_{Y|X}(y \mid x)}{\int_{-\infty}^{\infty} f_X(x) f_{Y|X}(y \mid x)\, dx}. \tag{33}$$

This formula is the continuous analog to the famous Bayes' rule discussed in Chapter 1.

In Chapters 3 and 4 we considered random variables X and Y which were both discrete. So far in Chapter 6 we have mainly considered random variables X and Y which are both continuous. There are cases when one is interested simultaneously in both discrete and continuous random variables. It should be clear to the reader how we could modify our discussion to include this possibility. Some of the most interesting applications of Formula (33) are of this type.

Example 9. Suppose the number of automobile accidents a driver will be involved in during a one-year period is a random variable Y having a Poisson distribution with parameter λ, where λ depends on the driver. If we choose a driver at random from some population, we can let λ vary and define a continuous random variable Λ having density f_Λ. The conditional density of Y given $\Lambda = \lambda$ is the Poisson density with parameter λ given by

$$f_{Y|\Lambda}(y \mid \lambda) = \begin{cases} \dfrac{\lambda^y e^{-\lambda}}{y!} & \text{for} \quad y = 0, 1, 2, \ldots, \\ 0, & \text{elsewhere.} \end{cases}$$

Thus the joint density of Λ and Y is

$$f(\lambda, y) = \begin{cases} \dfrac{f_\Lambda(\lambda)\lambda^y e^{-\lambda}}{y!} & \text{for} \quad y = 0, 1, 2, \ldots, \\ 0, & \text{elsewhere.} \end{cases}$$

In general we cannot find a nice formula for the marginal density of Y or the conditional density of Λ given $Y = y$, since we cannot evaluate the required integrals. We can find simple formulas, however, in the special case when f is a gamma density $\Gamma(\alpha, \beta)$, so that

$$f_\Lambda(\lambda) = \begin{cases} \dfrac{\beta^\alpha \lambda^{\alpha - 1} e^{-\lambda\beta}}{\Gamma(\alpha)} & \text{for} \quad \lambda > 0, \\ 0, & \text{elsewhere.} \end{cases}$$

In this case,

$$f_Y(y) = \int_{-\infty}^{\infty} f(\lambda, y)\, d\lambda$$

$$= \int_0^\infty \frac{\beta^\alpha \lambda^{\alpha-1} e^{-\lambda\beta}}{\Gamma(\alpha)} \frac{\lambda^y e^{-\lambda}}{y!} \, d\lambda$$

$$= \frac{\beta^\alpha}{y!\Gamma(\alpha)} \int_0^\infty \lambda^{\alpha+y-1} e^{-\lambda(\beta+1)} \, d\lambda$$

$$= \frac{\Gamma(\alpha + y)\beta^\alpha}{y!\Gamma(\alpha)(\beta + 1)^{\alpha+y}}.$$

The value of the last integral was obtained by using Formula (34) of Chapter 5. We leave it as an exercise for the reader to show that f_Y is the negative binomial density with parameters α and $p = \beta/(1 + \beta)$. We also have that for $\lambda > 0$ and y a nonnegative integer,

$$f_{\Lambda|Y}(\lambda \mid y) = \frac{f(\lambda, y)}{f_Y(y)}$$

$$= \frac{\beta^\alpha \lambda^{\alpha+y-1} e^{-\lambda(\beta+1)} y!\Gamma(\alpha)(\beta + 1)^{\alpha+y}}{\Gamma(\alpha) y!\Gamma(\alpha + y)\beta^\alpha}$$

$$= \frac{(\beta + 1)^{\alpha+y} \lambda^{\alpha+y-1} e^{-\lambda(\beta+1)}}{\Gamma(\alpha + y)},$$

which says that the conditional density of Λ given $Y = y$ is the gamma density $\Gamma(\alpha + y, \beta + 1)$. If someone in the insurance industry wanted to solve problems of this type he would quite possibly try to approximate the true density f_Λ by a gamma density $\Gamma(\alpha, \beta)$, where α and β are chosen to make the approximation as good as possible.

6.4. Properties of multivariate distributions

The concepts discussed so far in this chapter for two random variables X and Y are readily extended to n random variables. In this section we indicate briefly how this is done.

Let $X_1, \ldots, X_n$ be n random variables defined on a common probability space. Their *joint distribution function* F is defined by

$$F(x_1, \ldots, x_n) = P(X_1 \le x_1, \ldots, X_n \le x_n), \qquad -\infty < x_1, \ldots, x_n < \infty.$$

The marginal distribution functions F_{X_m}, $m = 1, \ldots, n$, are defined by

$$F_{X_m}(x_m) = P(X_m \le x_m), \qquad -\infty < x_m < \infty.$$

The value of $F_{X_m}(x_m)$ can be obtained from F by letting $x_1, \ldots, x_{m-1}, x_{m+1}, \ldots, x_n$, all approach $+\infty$.

A nonnegative function f is called a joint density function (with respect to integration) for the joint distribution function F, or for the random variables $X_1, \ldots, X_n$, if

(34) $$F(x_1, \ldots, x_n) = \int_{-\infty}^{x_1} \cdots \int_{-\infty}^{x_n} f(u_1, \ldots, u_n)\, du_1 \cdots du_n, \qquad -\infty < x_1, \ldots, x_n < \infty.$$

Under some further mild conditions the equation

$$f(x_1, \ldots, x_n) = \frac{\partial^n}{\partial x_1 \cdots \partial x_n} F(x_1, \ldots, x_n)$$

is valid at the continuity points of F. If (34) holds and A is any subset of R^n of the type considered in calculus, then

$$P((X_1, \ldots, X_n) \in A) = \int \cdots \int_A f(x_1, \ldots, x_n)\, dx_1 \cdots dx_n.$$

In particular

(35) $$\int_{-\infty}^{\infty} \cdots \int_{-\infty}^{\infty} f(x_1, \ldots, x_n)\, dx_1 \cdots dx_n = 1$$

and if $a_m \le b_m$ for $m = 1, \ldots, n$, then

$$P(a_1 \le X_1 \le b_1, \ldots, a_n \le X_n \le b_n) = \int_{a_1}^{b_1} \cdots \int_{a_n}^{b_n} f(x_1, \ldots, x_n)\, dx_1 \cdots dx_n.$$

The random variable X_m has the marginal density f_{X_m} obtained by integrating f over the remaining $n - 1$ variables. For example,

$$f_{X_2}(x_2) = \int_{-\infty}^{\infty} \cdots \int_{-\infty}^{\infty} f(x_1, \ldots, x_n)\, dx_1\, dx_3 \cdots dx_n.$$

In general, the random variables $X_1, \ldots, X_n$ are called independent if whenever $a_m \le b_m$ for $m = 1, \ldots, n$, then

$$P(a_1 < X_1 \le b_1, \ldots, a_n < X_n \le b_n) = P(a_1 < X_1 \le b_1) \cdots P(a_n < X_n \le b_n).$$

A necessary and sufficient condition for independence is that

$$F(x_1, \ldots, x_n) = F_{X_1}(x_1) \cdots F_{X_n}(x_n), \qquad -\infty < x_1, \ldots, x_n < \infty.$$

The necessity is obvious, but the sufficiency part for $n > 2$ is tricky and will not be proved here. If F has a density f, then $X_1, \ldots, X_n$ are independent if and only if f can be chosen so that

$$f(x_1, \ldots, x_n) = f_{X_1}(x_1) \cdots f_{X_n}(x_n), \qquad -\infty < x_1, \ldots, x_n < \infty.$$

One can also define an n-dimensional density directly as a nonnegative function on R^n such that (35) holds. The simplest way to construct n-dimensional densities is to start with n one-dimensional densities $f_1, \ldots, f_n$ and define f by

(36) $$f(x_1, \ldots, x_n) = f_1(x_1) \cdots f_n(x_n), \qquad -\infty < x_1, \ldots, x_n < \infty.$$

If $X_1, \ldots, X_n$ are random variables whose joint density f is given by (36), then $X_1, \ldots, X_n$ are independent and X_m has the marginal density f_m.

Example 10. Let $X_1, \ldots, X_n$ be independent random variables, each having an exponential density with parameter λ. Find the joint density of $X_1, \ldots, X_n$.

The density of X_m is given by

$$f_{X_m}(x_m) = \begin{cases} \lambda e^{-\lambda x_m} & \text{for} \quad 0 < x_m < \infty, \\ 0, & \text{elsewhere.} \end{cases}$$

Thus f is given by

$$f(x_1, \ldots, x_n) = \begin{cases} \lambda^n e^{-\lambda(x_1 + \cdots + x_n)} & \text{for} \quad x_1, \ldots, x_n > 0, \\ 0, & \text{elsewhere.} \end{cases}$$

In order to compute the density of the sum of n independent random variables, and for several other purposes, we need the following fact.

> ***Theorem 4*** *Let $X_1, \ldots, X_n$ be independent random variables. Let Y be a random variable defined in terms of $X_1, \ldots, X_m$, and let Z be a random variable defined in terms of $X_{m+1}, \ldots, X_n$ (where $1 \leq m < n$). Then Y and Z are independent.*

The proof of this theorem will not be given since it involves arguments from measure theory.

Using this theorem and an argument involving mathematical induction, we can extend Theorems 1 and 2 to sums of independent random variables, as follows.

> ***Theorem 5*** *Let $X_1, \ldots, X_n$ be independent random variables such that X_m has the gamma density $\Gamma(\alpha_m, \lambda)$ for $m = 1, \ldots, n$. Then $X_1 + \cdots + X_n$ has the gamma density $\Gamma(\alpha, \lambda)$, where*
>
> $$\alpha = \alpha_1 + \cdots + \alpha_n.$$

Recall that the exponential density with parameter λ is the same as the gamma density $\Gamma(1, \lambda)$. Thus as a special case of this theorem we have the following corollary: *If $X_1, \ldots, X_n$ are independent random variables, each having an exponential density with parameter λ, then $X_1 + \cdots + X_n$ has the gamma density $\Gamma(n, \lambda)$.*

> ***Theorem 6*** *Let $X_1, \ldots, X_n$ be independent random variables such that X_m has the normal density $n(\mu_m, \sigma_m^2)$, $m = 1, \ldots, n$. Then $X_1 + \cdots + X_n$ has the normal density $n(\mu, \sigma^2)$, where*
>
> $$\mu = \mu_1 + \cdots + \mu_n \quad \text{and} \quad \sigma^2 = \sigma_1^2 + \cdots + \sigma_n^2.$$

If $X_1, \ldots, X_n$ has a joint density f, then any subcollection of these random variables has a joint density which can be found by integrating over the remaining variables. For example, if $1 \leq m < n$,

$$f_{X_1,\ldots,X_m}(x_1, \ldots, x_m) = \int_{-\infty}^{\infty} \cdots \int_{-\infty}^{\infty} f(x_1, \ldots, x_n)\, dx_{m+1} \cdots dx_n.$$

The conditional density of a subcollection of $X_1, \ldots, X_n$ given the remaining variables can also be defined in an obvious manner. Thus the conditional density of $X_{m+1}, \ldots, X_n$ given $X_1, \ldots, X_m$ is defined by

$$f_{X_{m+1},\ldots,X_n|X_1,\ldots,X_m}(x_{m+1}, \ldots, x_n \mid x_1, \ldots, x_m) = \frac{f(x_1, \ldots, x_n)}{f_{X_1,\ldots,X_m}(x_1, \ldots, x_m)},$$

where f is the joint density of $X_1, \ldots, X_n$.
Often conditional densities are expressed in terms of a somewhat different notation. For example, let $n + 1$ random variables $X_1, \ldots, X_n, Y$ have joint density f. Then the conditional density of Y given $X_1, \ldots, X_n$ is defined by

$$f_{Y|X_1,\ldots,X_n}(y \mid x_1, \ldots, x_n) = \frac{f(x_1, \ldots, x_n, y)}{f_{X_1,\ldots,X_n}(x_1, \ldots, x_n)}.$$

6.5. Order statistics*

Let $U_1, \ldots, U_n$ be independent continuous random variables, each having distribution function F and density function f. Let $X_1, \ldots, X_n$ be random variables obtained by letting $X_1(\omega), \ldots, X_n(\omega)$ be the set $U_1(\omega), \ldots, U_n(\omega)$ permuted so as to be in increasing order. In particular, X_1 and X_n are defined to be the functions

$$X_1(\omega) = \min(U_1(\omega), \ldots, U_n(\omega))$$

and

$$X_n(\omega) = \max(U_1(\omega), \ldots, U_n(\omega)).$$

The random variable X_k is called the kth *order statistic*. Another related variable of interest is the range R, defined by

$$\begin{aligned} R(\omega) &= X_n(\omega) - X_1(\omega) \\ &= \max(U_1(\omega), \ldots, U_n(\omega)) - \min(U_1(\omega), \ldots, U_n(\omega)). \end{aligned}$$

It follows from the assumptions on $U_1, \ldots, U_n$ that, with probability one, the U_i's are distinct and hence $X_1 < X_2 < \cdots < X_n$.

To illustrate these definitions numerically, suppose $U_1(\omega) = 4.8$, $U_2(\omega) = 3.5$, and $U_3(\omega) = 4.3$. Then $X_1(\omega) = 3.5$, $X_2(\omega) = 4.3$, $X_3(\omega) = 4.8$, and $R(\omega) = 1.3$.

Example 11. Consider a machine having n parts whose failure times $U_1, \ldots, U_n$ satisfy the assumptions of this section. Then X_k is the time it takes for k of the parts to fail. If the entire machine fails as soon as a single part fails, then $X_1 = \min(U_1, \ldots, U_n)$ is the failure time of the machine. If the machine does not fail until all its parts have failed, then $X_n = \max(U_1, \ldots, U_n)$ is the failure time of the machine.

Example 12. Let n hopefully identical parts be manufactured in a single run of an assembly line and let $U_1, \ldots, U_n$ denote the lengths of the n parts. An inspector might be interested in the minimum length X_1 and maximum length X_n to check if they are within certain tolerance limits. If the parts are to be interchangeable the amount of variation in the lengths may have to be kept small. One possible measure of this variation is the range R of the lengths.

We will now compute the distribution function of the kth order statistic X_k. Let $-\infty < x < \infty$. The probability that exactly j of the U_i's lie in $(-\infty, x]$ and $(n - j)$ lie in (x, ∞) is

$$\binom{n}{j} F^j(x)(1 - F(x))^{n-j},$$

because the binomial distribution with parameters n and $p = F(x)$ is applicable. The event $\{X_k \le x\}$ occurs if and only if k or more of the U_i's lie in $(-\infty, x]$. Thus

$$\begin{aligned} F_{X_k}(x) &= P(X_k \le x) \\ &= \sum_{j=k}^{n} \binom{n}{j} F^j(x)(1 - F(x))^{n-j}, \qquad -\infty < x < \infty. \end{aligned} \tag{37}$$

In particular the distribution functions of X_n and X_1 can be written very simply as

$$F_{X_n}(x) = (F(x))^n, \qquad -\infty < x < \infty,$$

and

$$F_{X_1}(x) = 1 - (1 - F(x))^n, \qquad -\infty < x < \infty.$$

In order to find the corresponding density functions, we must differentiate these quantities. We easily find that

$$f_{X_n}(x) = nf(x)F^{n-1}(x), \qquad -\infty < x < \infty,$$

and

$$f_{X_1}(x) = nf(x)(1 - F(x))^{n-1}, \qquad -\infty < x < \infty.$$

The corresponding derivation for X_k in general is slightly more complicated. From (37),

$$\begin{aligned} f_{X_k}(x) &= \sum_{j=k}^{n} \frac{n!}{(j-1)!\,(n-j)!} f(x)F^{j-1}(x)(1 - F(x))^{n-j} \\ &\quad - \sum_{j=k}^{n-1} \frac{n!}{j!\,(n-j-1)!} f(x)F^{j}(x)(1 - F(x))^{n-j-1} \\ &= \sum_{j=k}^{n} \frac{n!}{(j-1)!\,(n-j)!} f(x)F^{j-1}(x)(1 - F(x))^{n-j} \\ &\quad - \sum_{j=k+1}^{n} \frac{n!}{(j-1)!\,(n-j)!} f(x)F^{j-1}(x)(1 - F(x))^{n-j} \end{aligned}$$

and by cancellation

$$(38)\qquad f_{X_k}(x) = \frac{n!}{(k-1)!\,(n-k)!} f(x)F^{k-1}(x)(1 - F(x))^{n-k}, \qquad -\infty < x < \infty.$$

In order to find the density of the range R we will first find the joint density of X_1 and X_n. We assume that $n \geq 2$ (since $R = 0$ if $n = 1$). Let $x \leq y$. Then

$$\begin{aligned} P(X_1 > x, X_n \leq y) &= P(x < U_1 \leq y, \ldots, x < U_n \leq y) \\ &= (F(y) - F(x))^n, \end{aligned}$$

and of course

$$P(X_n \leq y) = F^n(y).$$

Consequently

$$\begin{aligned} F_{X_1,X_n}(x, y) &= P(X_1 \leq x, X_n \leq y) \\ &= P(X_n \leq y) - P(X_1 > x, X_n \leq y) \\ &= F^n(y) - (F(y) - F(x))^n. \end{aligned}$$

The joint density is given by

$$\begin{aligned} f_{X_1,X_n}(x, y) &= \frac{\partial^2}{\partial x\,\partial y} F_{X_1,X_n}(x, y) \\ &= n(n-1)f(x)f(y)(F(y) - F(x))^{n-2}, \qquad x \leq y. \end{aligned}$$

It is obvious and easily shown that

$$f_{X_1,X_n}(x, y) = 0, \qquad x > y.$$

By slightly modifying the argument used in Section 6.2.1 to find the density of a sum, we find that the density of $R = X_n - X_1$ is given by

$$f_R(r) = \int_{-\infty}^{\infty} f_{X_1,X_n}(x, r + x)\,dx.$$

In other words

$$f_R(r) = \begin{cases} n(n-1) \displaystyle\int_{-\infty}^{\infty} f(x)f(r + x)(F(r + x) - F(x))^{n-2}\,dx, & r > 0 \\ 0, & r < 0. \end{cases}$$

These formulas can all be evaluated simply when $U_1, \ldots, U_n$ are independent and uniformly distributed in $(0, 1)$. This is left as an exercise.

There is a "heuristic" way for deriving these formulas which is quite helpful. We will illustrate it by rederiving the formula for f_{X_k}. Let dx denote a small positive number. Then we have the approximation

$$f_{X_k}(x)\,dx \approx P(x \leq X_k \leq x + dx).$$

The most likely way for the event $\{x \le X_k \le x + dx\}$ to occur is that $k - 1$ of the U_i's should lie in $(-\infty, x]$, one of the U_i's should lie in $(x, x + dx]$, and $n - k$ of the U_i's should lie in $(x + dx, \infty)$ (see Figure 5). The derivation of the multinomial distribution given in Chapter 3 is applicable and the probability that the indicated number of the U_i's will lie in the appropriate intervals is

$$f_{X_k}(x)\,dx \approx \frac{n!}{(k-1)!\,1!\,(n-k)!} \times \left(\int_{-\infty}^{x} f(u)\,du\right)^{k-1} \int_{x}^{x+dx} f(u)\,du \left(\int_{x+dx}^{\infty} f(u)\,du\right)^{n-k}$$

$$\approx \frac{n!}{(k-1)!\,(n-k)!} f(x)\,dx F^{k-1}(x)(1 - F(x))^{n-k},$$

from which we get (38). We shall not attempt to make this method rigorous.

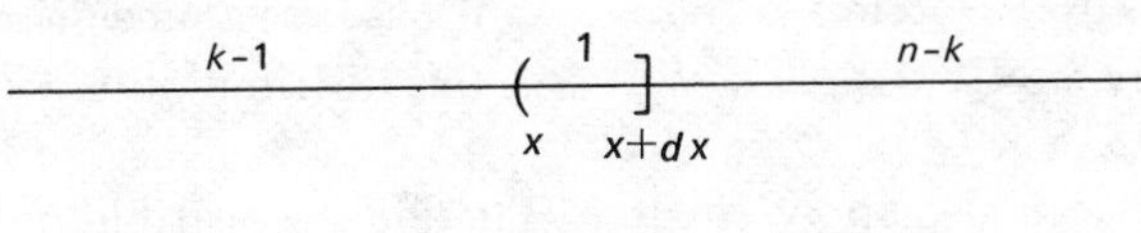

Figure 5

6.6. Sampling distributions*

Let $X_1, \ldots, X_n$ be independent random variables, each having the normal density $n(0, \sigma^2)$. In this section we will find the distribution functions of several random variables defined in terms of the X's. Besides providing applications of the preceding material, these distribution functions are of fundamental importance in statistical inference, and will be needed in Volume II.

The constant σ^2 is convenient but unessential since $X_1/\sigma, \ldots, X_n/\sigma$ are independent and each has the standard normal density $n(0, 1)$. Thus we could always take $\sigma^2 = 1$ with no loss of generality.

By Theorem 6 the random variable $X_1 + \cdots + X_n$ has the normal density with parameters 0 and $n\sigma^2$. If we divide this sum by various constants we can get alternative forms of this result. Thus

$$\frac{X_1 + \cdots + X_n}{n}$$

is normally distributed with parameters 0 and σ^2/n, and

$$\frac{X_1 + \cdots + X_n}{\sigma\sqrt{n}}$$

has the standard normal density $n(0, 1)$.

Since X_1/σ has the standard normal density, it follows from Example 12 of Chapter 5 that X_1^2/σ^2 has the gamma density $\Gamma(1/2, 1/2)$. Thus by Theorem 5

$$\frac{X_1^2 + \cdots + X_n^2}{\sigma^2}$$

has the gamma density $\Gamma(n/2, 1/2)$. This particular gamma density is very important in statistics. There the corresponding random variable is said to have a chi-square (χ^2) distribution with n degrees of freedom, denoted by $\chi^2(n)$. By applying Theorem 5 we will obtain the following result about χ^2 distributions.

Theorem 7 *Let $Y_1, \ldots, Y_n$ be independent random variables such that Y_m has the $\chi^2(k_m)$ distribution. Then $Y_1 + \cdots + Y_n$ has the $\chi^2(k)$ distribution, where $k = k_1 + \cdots + k_n$.*

Proof. By assumption, Y_m has the gamma distribution $\Gamma(k_m/2, 1/2)$. Thus by Theorem 5, $Y_1 + \cdots + Y_n$ has the gamma distribution $\Gamma(k/2, 1/2)$ where $k = k_1 + \cdots + k_n$. But this distribution is $\chi^2(k)$ by definition. ∎

We can also apply Theorem 3 to find the distribution of the ratio of two independent random variables Y_1 and Y_2 having distributions $\chi^2(k_1)$ and $\chi^2(k_2)$ respectively. It is traditional in statistics to express the results in terms of the normalized variables Y_1/k_1 and Y_2/k_2. The distribution of

$$\frac{Y_1/k_1}{Y_2/k_2}$$

is known as the F distribution with k_1 and k_2 degrees of freedom, denoted by $F(k_1, k_2)$.

Theorem 8 *Let Y_1 and Y_2 be independent random variables having distributions $\chi^2(k_1)$ and $\chi^2(k_2)$. Then the random variable*

$$\frac{Y_1/k_1}{Y_2/k_2},$$

which has the distribution $F(k_1, k_2)$, has the density f given by $f(x) = 0$ for $x \le 0$ and

$$(39) \quad f(x) = \frac{(k_1/k_2)\,\Gamma[(k_1 + k_2)/2]\,(k_1x/k_2)^{(k_1/2)-1}}{\Gamma(k_1/2)\,\Gamma(k_2/2)\,[1 + (k_1x/k_2)]^{(k_1+k_2)/2}}, \qquad x > 0.$$

Proof. By Theorem 3, the random variable Y_1/Y_2 has density g, where g is given by (24) with $\alpha_1 = k_1/2$ and $\alpha_2 = k_2/2$. Thus the density f of k_2Y_1/k_1Y_2 is given by

$$f(x) = \frac{k_1}{k_2}\, g\left(\frac{k_1x}{k_2}\right)$$

and (39) now follows from (24). ∎

We can apply this result to the random variables $X_1, \ldots, X_n$ defined at the beginning of this section. Let $1 \le m < n$. By Theorem 4, the random variables

$$\frac{X_1^2 + \cdots + X_m^2}{\sigma^2} \quad \text{and} \quad \frac{X_{m+1}^2 + \cdots + X_n^2}{\sigma^2}$$

are independent. Since they have the respective distributions $\chi^2(m)$ and $\chi^2(n - m)$ we see that the random variable

$$\frac{(X_1^2 + \cdots + X_m^2)/m}{(X_{m+1}^2 + \cdots + X_n^2)/(n - m)}$$

has the $F(m, n - m)$ distribution and the density given by (39), where $k_1 = m$ and $k_2 = n - m$. Tables of F distributions are given in Volume II.

The case $m = 1$ is especially important. The random variable

$$\frac{X_1^2}{(X_2^2 + \cdots + X_n^2)/(n - 1)}$$

has the $F(1, n - 1)$ distribution. We can use this fact to find the distribution of

$$Y = \frac{X_1}{\sqrt{(X_2^2 + \cdots + X_n^2)/(n - 1)}}.$$

Since X_1 has a symmetric density function and is independent of the random variable $\sqrt{(X_2^2 + \cdots + X_n^2)/(n - 1)}$, it follows easily from Theorem 2 of Chapter 5 that Y has a symmetric density function f_Y. By Example 5 of Chapter 5 the density f_{Y^2} is related to f_Y by

$$f_{Y^2}(z) = \frac{1}{2\sqrt{z}}(f_Y(-\sqrt{z}) + f_Y(\sqrt{z})), \qquad z > 0.$$

By using the symmetry of f_Y and letting $z = y^2$ we see that

$$f_Y(y) = |y| f_{Y^2}(y^2).$$

Since Y^2 has the $F(1, n - 1)$ density given by (39) with $k_1 = 1$ and $k_2 = k = n - 1$, we now find that

$$f_Y(y) = \frac{|y|\,(1/k)\,\Gamma[(k + 1)/2]\,(y^2/k)^{-1/2}}{\Gamma(1/2)\,\Gamma(k/2)\,[1 + (y^2/k)]^{(k+1)/2}}.$$

Since $\Gamma(1/2) = \sqrt{\pi}$, this expression reduces to

$$(40) \qquad f_Y(y) = \frac{\Gamma[(k + 1)/2]\,[1 + (y^2/k)]^{-(k+1)/2}}{\sqrt{k\pi}\,\Gamma(k/2)}, \qquad -\infty < y < \infty.$$

A random variable whose density is given by (40) is said to have a t distribution with k degrees of freedom. We observe that the t distribution with 1 degree of freedom is the same as the Cauchy distribution discussed in Chapter 4. Tables of t distributions are given in Volume II.

The distribution of the random variable

$$Y = \frac{X_1}{\sqrt{(X_2^2 + \cdots + X_n^2)/(n-1)}},$$

which is a t distribution with $n - 1$ degrees of freedom, depends only on the fact that

$$\frac{X_1}{\sigma} \quad \text{and} \quad \frac{X_2^2 + \cdots + X_n^2}{\sigma^2}$$

are independent and distributed respectively as $n(0, 1)$ and $\chi^2(n - 1)$. Thus we have the following result.

Theorem 9 *Let X and Y be independent variables with the respective distributions $n(0, 1)$ and $\chi^2(k)$. Then*

$$\frac{X}{\sqrt{Y/k}}$$

has a t distribution with k degrees of freedom.

6.7. Multidimensional changes of variables*

Let $X_1, \ldots, X_n$ be continuous random variables having joint density f. Let $Y_1, \ldots, Y_n$ be random variables defined in terms of the X's. In this section we will discuss a method for finding the joint density of the Y's in terms of f. We will consider mainly the case when the Y's are defined as linear functions of the X's.

Suppose then that

$$Y_i = \sum_{j=1}^{n} a_{ij}X_j, \qquad i = 1, \ldots, n.$$

The constant coefficients a_{ij} determine an $n \times n$ matrix

$$A = [a_{ij}] = \begin{bmatrix} a_{11} & \cdots & a_{1n} \\ \vdots & & \vdots \\ a_{n1} & \cdots & a_{nn} \end{bmatrix}.$$

Associated with such a matrix is its determinant

$$\det A = \begin{vmatrix} a_{11} & \cdots & a_{11} \\ \vdots & & \vdots \\ a_{n1} & \cdots & a_{nn} \end{vmatrix}.$$

If $\det A \neq 0$ there is a unique inverse matrix $B = [b_{ij}]$ such that $BA = I$ or equivalently

$$\sum_{k=1}^{n} b_{ik}a_{kj} = \begin{cases} 1, & i = j, \\ 0, & i \neq j. \end{cases} \tag{41}$$

The constants b_{ij} can be obtained by solving for each i the system (41) of n equations in the n unknowns $b_{i1}, \ldots, b_{in}$. Alternatively, the constants b_{ij} are uniquely defined by requiring that the equations

$$y_i = \sum_{j=1}^{n} a_{ij}x_j, \qquad i = 1, \ldots, n,$$

have solutions

$$x_i = \sum_{j=1}^{n} b_{ij}y_j, \qquad i = 1, \ldots, n. \tag{42}$$

Theorem 10 *Let $X_1, \ldots, X_n$ be continuous random variables having joint density f and let random variables $Y_1, \ldots, Y_n$ be defined by*

$$Y_i = \sum_{j=1}^{n} a_{ij}X_j, \qquad i = 1, \ldots, n,$$

where the matrix $A = [a_{ij}]$ has nonzero determinant det A. *Then $Y_1, \ldots, Y_n$ have joint density $f_{Y_1,\ldots,Y_n}$ given by*

$$f_{Y_1,\ldots,Y_n}(y_1, \ldots, y_n) = \frac{1}{|\det A|} f(x_1, \ldots, x_n), \tag{43}$$

where the x's are defined in terms of the y's by (42) *or as the unique solution to the equations $y_i = \sum_{j=1}^{n} a_{ij}x_j$.*

This theorem, which we will not prove here, is equivalent to a theorem proved in advanced calculus courses in a more general setting involving "Jacobians." From the general result proved in advanced calculus, we can extend the above theorem to nonlinear changes of variables. We will describe this extension briefly, although it will not be needed later.

Let the Y's be defined in terms of the X's by

$$Y_i = g_i(X_1, \ldots, X_n), \qquad i = 1, \ldots, n.$$

Consider the corresponding equations

$$y_i = g_i(x_1, \ldots, x_n), \qquad i = 1, \ldots, n. \tag{44}$$

Suppose that these equations define the x's uniquely in terms of the y's, that the partial derivatives $\partial y_i/\partial x_j$ exist and are continuous, and that the Jacobian

$$J(x_1, \ldots, x_n) = \begin{vmatrix} \dfrac{\partial y_1}{\partial x_1} & \cdots & \dfrac{\partial y_1}{\partial x_n} \\ \vdots & & \vdots \\ \dfrac{\partial y_n}{\partial x_1} & \cdots & \dfrac{\partial y_n}{\partial x_n} \end{vmatrix}$$

is everywhere nonzero. Then the random variables $Y_1, \ldots, Y_n$ are continuous and have a joint density given by

$$(45) \qquad f_{Y_1,\ldots,Y_n}(y_1, \ldots, y_n) = \frac{1}{|J(x_1, \ldots, x_n)|} f(x_1, \ldots, x_n),$$

where the x's are defined implicitly in terms of the y's by (44). This change of variable formula can be extended still further by requiring that the functions g_i be defined only on some open subset S of R^n such that

$$P((X_1, \ldots, X_n) \in S) = 1.$$

In the special case when $y_i = \sum_{j=1}^{n} a_{ij}x_j$, we see that $\partial y_i/\partial x_j = a_{ij}$ and $J(x_1, \ldots, x_n)$ is just the constant $\det A = \det [a_{ij}]$. So it is clear that (45) reduces to (43) in the linear case.

Example 13. Let $X_1, \ldots, X_n$ be independent random variables each having an exponential density with parameter λ. Define $Y_1, \ldots, Y_n$ by $Y_i = X_1 + \cdots + X_i$, $1 \le i \le n$. Find the joint density of $Y_1, \ldots, Y_n$.

The matrix $[a_{ij}]$ is

$$\begin{bmatrix} 1 & 0 & \cdot & \cdot & \cdot & 0 \\ 1 & 1 & 0 & & & \vdots \\ \vdots & & & & & 0 \\ 1 & 1 & \cdot & \cdot & \cdot & 1 \end{bmatrix}$$

Its determinant is clearly 1. The equations

$$y_i = x_1 + \cdots + x_i, \qquad i = 1, \ldots, n,$$

have the solution

$$x_1 = y_1,$$
$$x_i = y_i - y_{i-1}, \qquad i = 2, \ldots, n.$$

The joint density of $X_1, \ldots, X_n$ is given by

$$(46) \qquad f(x_1, \ldots, x_n) = \begin{cases} \lambda^n e^{-\lambda(x_1 + \cdots + x_n)}, & x_1, \ldots, x_n > 0, \\ 0, & \text{elsewhere.} \end{cases}$$

Thus the joint density $f_{Y_1,\ldots,Y_n}$ is given by

$$(47) \qquad f_{Y_1,\ldots,Y_n}(y_1, \ldots, y_n) = \begin{cases} \lambda^n e^{-\lambda y_n}, & 0 < y_1 < \cdots < y_n, \\ 0, & \text{elsewhere.} \end{cases}$$

Of course, one can apply the theorem in the reverse direction. Thus if $Y_1, \ldots, Y_n$ have the joint density given by (47), and random variables $X_1, \ldots, X_n$ are defined by $X_1 = Y_1$ and $X_i = Y_i - Y_{i-1}$ for $2 \le i \le n$, then the X's have the joint density f given by (46). In other words, $X_1, \ldots, X_n$ are independent and each has an exponential distribution with parameter λ. This result will be used in Chapter 9 in connection with Poisson processes.

Exercises

1 Let X and Y be continuous random variables having joint density function f. Find the joint distribution function and the joint density function of the random variables $W = a + bX$ and $Z = c + dY$, where $b > 0$ and $d > 0$. Show that if X and Y are independent, then W and Z are independent.

2 Let X and Y be continuous random variables having joint distribution function F and joint density function f. Find the joint distribution function and joint density function of the random variables $W = X^2$ and $Z = Y^2$. Show that if X and Y are independent, then W and Z are independent.

3 Let X and Y be independent random variables each uniformly distributed on $(0, 1)$. Find
(a) $P(|X - Y| \leq .5)$,
(b) $P\left(\left|\frac{X}{Y} - 1\right| \leq .5\right)$,
(c) $P(Y \geq X \mid Y \geq 1/2)$.

4 Let X and Y be independent random variables each having the normal density $n(0, \sigma^2)$. Find $P(X^2 + Y^2 \leq 1)$. *Hint:* Use polar coordinates.

5 Let X and Y have a joint density f that is uniform over the interior of the triangle with vertices at $(0, 0)$, $(2, 0)$, and $(1, 2)$. Find $P(X \leq 1$ and $Y \leq 1)$.

6 Suppose the times it takes two students to solve a problem are independently and exponentially distributed with parameter λ. Find the probability that the first student will take at least twice as long as the second student to solve the problem.

7 Let X and Y be continuous random variables having joint density f given by $f(x, y) = \lambda^2 e^{-\lambda y}$, $0 \leq x \leq y$, and $f(x, y) = 0$ elsewhere. Find the marginal densities of X and Y. Find the joint distribution function of X and Y.

8 Let $f(x, y) = c(y - x)^\alpha$, $0 \leq x < y \leq 1$, and $f(x, y) = 0$ elsewhere.
(a) For what values of α can c be chosen to make f a density function?
(b) How should c be chosen (when possible) to make f a density?
(c) Find the marginal densities of f.

9 Let $f(x, y) = ce^{-(x^2 - xy + 4y^2)/2}$, $-\infty < x, y < \infty$. How should c be chosen to make f a density? Find the marginal densities of f.

10 Let X and Y be independent continuous random variables having joint density f. Derive a formula for the density of $Z = Y - X$.

11 Let X and Y be independent continuous random variables having the indicated marginal densities. Find the density of $Z = X + Y$.

(a) X and Y are exponentially distributed with parameters λ_1 and λ_2, where $\lambda_1 \neq \lambda_2$.

(b) X is uniform on $(0, 1)$, and Y is exponentially distributed with parameter λ.

12 Let X and Y have a joint density f given in Exercise 8. Find the density of $Z = X + Y$.

13 Let X and Y be independent and uniformly distributed on (a, b). Find the density of $Z = |Y - X|$.

14 Let X and Y be continuous random variables having joint density f. Derive a formula for the density of $Z = aX + bY$, where $b \neq 0$.

15 Let f be a Beta density with parameters $\alpha_1 > 1$ and $\alpha_2 > 1$. Where does f take on its maximum value?

16 Let X and Y be independent random variables having respective normal densities $n(\mu_1, \sigma_1^2)$ and $n(\mu_2, \sigma_2^2)$. Find the density of $Z = Y - X$.

17 Let a point be chosen randomly in the plane in such a manner that its x and y coordinates are independently distributed according to the normal density $n(0, \sigma^2)$. Find the density function for the random variable R denoting the distance from the point to the origin. (This density occurs in electrical engineering and is known there as a *Rayleigh* density.)

18 Let X and Y be continuous random variables having joint density f. Derive a formula for the density of $Z = XY$.

19 Let X and Y be independent random variables each having the normal density $n(0, \sigma^2)$. Show that Y/X and $Y/|X|$ both have the Cauchy density.

20 Let X and Y be as in Exercise 19. Find the density of $Z = |Y|/|X|$.

21 Let X and Y be independent random variables each having an exponential distribution with parameter λ. Find the density of $Z = Y/X$.

22 Let X and Y be independent random variables having respective gamma densities $\Gamma(\alpha_1, \lambda)$ and $\Gamma(\alpha_2, \lambda)$. Find the density of $Z = X/(X + Y)$. *Hint:* Express Z in terms of Y/X.

23 Let X and Y have joint density f as indicated below. Find the conditional density $f_{Y|X}$ in each case:

(a) f as in Exercise 7,

(b) f as in Exercise 8,

(c) f as in Exercise 9.

24 Let X and Y be distributed as in Example 7. Find $P(Y \leq 2 \mid X = 1)$.

25 Show that the marginal density f_Y in Example 9 is negative binomial with parameters α and $p = \beta/(\beta + 1)$. *Hint:* Use Formula (36) of Chapter 5.

26 Let Y be a discrete random variable having a binomial distribution with parameters n and p. Suppose now that p varies as random variable

Π having a Beta density with parameters α_1 and α_2. Find the conditional density of Π given $Y = y$.

27 Let Y be exponentially distributed with parameter λ. Let λ vary as a random variable Λ having the gamma density $\Gamma(\alpha, \beta)$. Find the marginal density of Y and the conditional density of Λ given $Y = y$.

28 Let X_1, X_2, X_3 denote the three components of the velocity of a molecule of gas. Suppose that X_1, X_2, X_3 are independent and each has the normal density $n(0, \sigma^2)$. In physics the magnitude of the velocity $Y = (X_1^2 + X_2^2 + X_3^2)^{1/2}$ is said to have a *Maxwell* distribution. Find f_Y.

29 Let $X_1, \ldots, X_n$ be independent random variables having a common normal density. Show that there are constants A_n and B_n such that

$$\frac{X_1 + \cdots + X_n - A_n}{B_n}$$

has the same density as X_1.

30 Let X_1, X_2, X_3 be independent random variables each uniformly distributed on $(0, 1)$. Find the density of the random variable $Y = X_1 + X_2 + X_3$. Find $P(X_1 + X_2 + X_3 \leq 2)$.

31 Let X_1 be chosen uniformly on $(0, 1)$, let X_2 be chosen uniformly on $(0, X_1)$, and let X_3 be chosen uniformly on $(0, X_2)$. Find the joint density of X_1, X_2, X_3 and the marginal density of X_3.

32 Let $U_1, \ldots, U_n$ be independent random variables each uniformly distributed over $(0, 1)$. Let X_k, $k = 1, \ldots, n$, and R be as in Section 6.5.
(a) Find the joint density of X_1 and X_n.
(b) Find the density of R.
(c) Find the density of X_k.

33 Let $U_1, \ldots, U_n$ be independent random variables each having an exponential density with parameter λ. Find the density of $X_1 = \min(U_1, \ldots, U_n)$.

34 Find a formula for the $\chi^2(n)$ density.

35 Let X and Y be independent random variables distributed respectively as $\chi^2(m)$ and $\chi^2(n)$. Find the density of $Z = X/(X + Y)$. *Hint:* Use the answer to Exercise 22.

36 Let X and Y be independent random variables each having the standard normal density. Find the joint density of $aX + bY$ and $bX - aY$, where $a^2 + b^2 > 0$. Use this to give another derivation of Theorem 2.

37 Let X and Y be independent random variables each having density f. Find the joint density of X and $Z = X + Y$.

38 Let X and Y be independent random variables each having an exponential density with parameter λ. Find the conditional density of X given $Z = X + Y = z$. *Hint:* Use the result of Exercise 37.

39 Solve Exercise 38 if X and Y are uniformly distributed on $(0, c)$.

40 Let U and V be independent random variables each having the standard normal density. Set $Z = \rho U + \sqrt{1 - \rho^2}\, V$, where $-1 < \rho < 1$.
(a) Find the density of Z.
(b) Find the joint density of U and Z.
(c) Find the joint density of $X = \mu_1 + \sigma_1 U$ and $Y = \mu_2 + \sigma_2 Z$, where $\sigma_1 > 0$ and $\sigma_2 > 0$. This joint density is known as a *bivariate normal* density.
(d) Find the conditional density of Y given $X = x$.

41 Let X and Y be positive continuous random variables having joint density f. Set $W = Y/X$ and $Z = X + Y$. Find the joint density of W and Z in terms of f. *Hint:* Use Equation (45).

42 Let X and Y be independent random variables having the respective gamma densities $\Gamma(\alpha_1, \lambda)$ and $\Gamma(\alpha_2, \lambda)$. Use Exercise 41 to show that Y/X and $X + Y$ are independent random variables.

43 Let R and Θ be independent random variables such that R has the Rayleigh density

$$f_R(r) = \begin{cases} \sigma^{-2} r\, e^{-r^2/2\sigma^2}, & r \geq 0, \\ 0, & r < 0, \end{cases}$$

and Θ is uniformly distributed on $(-\pi, \pi)$. Show that $X = R \cos \Theta$ and $Y = R \sin \Theta$ are independent random variables and that each has the normal density $n(0, \sigma^2)$. *Hint:* Use Equation (45).

7 Expectations and the Central Limit Theorem

In the first four sections of this chapter we extend the definition and properties of expectations to random variables which are not necessarily discrete. In Section 7.5 we discuss the Central Limit Theorem. This theorem, one of the most important in probability theory, justifies the approximation of many distribution functions by the appropriate normal distribution function.

7.1. Expectations of continuous random variables

Let us recall from Chapter 4 our definition of the expectation of a discrete random variable X having density f. We say that X has finite expectation if $\sum_x |x| f(x) < \infty$, and in that case we define its expectation EX as

$$EX = \sum_x x f(x).$$

The easiest way to define expectations of continuous random variables having densities is by analogy with the discrete case.

Definition 1 *Let X be a continuous random variable having density f. We say that X has finite expectation if*

$$\int_{-\infty}^{\infty} |x| f(x)\, dx < \infty,$$

and in that case we define its expectation by

$$EX = \int_{-\infty}^{\infty} x f(x)\, dx.$$

Using this definition we can easily calculate expectations of continuous random variables having the various densities discussed in Chapters 5 and 6.

Example 1. Let X be uniformly distributed on (a, b). Then

$$EX = \int_a^b x \left(\frac{1}{b-a}\right) dx = \left(\frac{1}{b-a}\right) \frac{x^2}{2}\bigg|_a^b = \frac{a+b}{2}.$$

Example 2. Let X have the gamma density $\Gamma(\alpha, \lambda)$. Then

$$\begin{aligned} EX &= \int_0^\infty x \frac{\lambda^\alpha}{\Gamma(\alpha)} x^{\alpha-1} e^{-\lambda x}\, dx \\ &= \frac{\lambda^\alpha}{\Gamma(\alpha)} \int_0^\infty x^\alpha e^{-\lambda x}\, dx \\ &= \frac{\lambda^\alpha}{\Gamma(\alpha)} \frac{\Gamma(\alpha+1)}{\lambda^{\alpha+1}} \\ &= \frac{\alpha}{\lambda}, \end{aligned}$$

where we have used Formulas (34) and (36) of Chapter 5. By setting $\alpha = 1$ we see that if X has an exponential density with parameter λ, then $EX = \lambda^{-1}$.

Example 3. Let X have the Cauchy density f given by

$$f(x) = \frac{1}{\pi(1 + x^2)}, \qquad -\infty < x < \infty.$$

Then X does not have finite expectation. For

$$\begin{aligned} \int_{-\infty}^\infty |x| \frac{1}{\pi(1+x^2)}\, dx &= \frac{2}{\pi} \int_0^\infty \frac{x}{1+x^2}\, dx \\ &= \frac{2}{\pi} \lim_{c\to\infty} \int_0^c \frac{x}{1+x^2}\, dx \\ &= \frac{1}{\pi} \lim_{c\to\infty} \log(1+x^2)\Big|_0^c \\ &= \infty. \end{aligned}$$

7.2. A general definition of expectation

The definition of expectation given in Section 7.1 is certainly appropriate from a computational point of view for the case of continuous random variables having densities. In order to define expectation in general, however, it is better to extend the notion of expectation directly from the discrete case to the general case. We will present only the basic ideas that motivate the general definition of expectation. The precise details require further background in the theory of measure and integration. We will assume in our discussion that all random variables under consideration are defined on a fixed probability space $(\Omega, \mathscr{A}, P)$.

Let X and Y be discrete random variables such that, for some $\varepsilon > 0$, $P(|X - Y| \le \varepsilon) = 1$. It follows from Theorems 2(iii) and 3 of Chapter 4 that if Y has finite expectation, then X has finite expectation and $|EX - EY| \leq \varepsilon$. It also follows that if Y does not have finite expectation, then neither does X. When expectation is defined in general these properties should continue to hold.

Let us assume this to be the case and let X be any random variable. Suppose we want to calculate EX with an error of at most ε, for some $\varepsilon > 0$. All we need to do is find a discrete random variable Y such that $P(|X - Y| \leq \varepsilon) = 1$ and calculate EY according to the methods of Chapter 4.

It is easy to find such approximations to X. Let X_ε be the discrete random variable defined by

$$(1) \qquad X_\varepsilon = \varepsilon k \quad \text{if} \quad \varepsilon k \leq X < \varepsilon(k + 1) \qquad \text{for the integer } k.$$

This random variable can also be defined in terms of the greatest integer function [ ] as $X_\varepsilon = \varepsilon[X/\varepsilon]$. If $\varepsilon = 10^{-n}$ for some nonnegative integer n, then $X_\varepsilon(\omega)$ can be obtained from $X(\omega)$ by writing $X(\omega)$ in decimal form and dropping all digits n or more places beyond the decimal point. It follows immediately from (1) that

$$X(\omega) - \varepsilon < X_\varepsilon(\omega) \leq X(\omega), \qquad \omega \in \Omega,$$

and hence $P(|X - X_\varepsilon| \leq \varepsilon) = 1$. The density function of X_ε is given by

$$f_{X_\varepsilon}(x) = \begin{cases} P(\varepsilon k \leq X < \varepsilon(k + 1)) & \text{if } x = \varepsilon k \text{ for the integer } k, \\ 0 & \text{elsewhere.} \end{cases}$$

The random variable X_ε has finite expectation if and only if

$$\sum_x |x| f_{X_\varepsilon}(x) = \sum_k |\varepsilon k| P(\varepsilon k \leq X < \varepsilon(k + 1)) < \infty,$$

in which case

$$EX_\varepsilon = \sum_k \varepsilon k P(\varepsilon k \leq X < \varepsilon(k + 1)).$$

These expressions can be written in terms of F_X. For

$$P(\varepsilon k \leq X < \varepsilon(k + 1)) = P(X < \varepsilon(k + 1)) - P(X < \varepsilon k)$$

and by Equation (5) of Chapter 5, $P(X < x) = F(x-)$ holds for all x. The following theorem, which we state without proof, will be used to give a general definition of expectation.

Theorem 1 *Let X be a random variable and let X_ε, $\varepsilon > 0$, be defined by* (1). *If X_ε has finite expectation for some $\varepsilon > 0$, then X_ε has finite expectation for all $\varepsilon > 0$ and*

$$\lim_{\varepsilon \to \infty} EX_\varepsilon$$

exists and is finite.

This theorem and our preceding discussion suggest the following general definition of expectation.

Definition 2 *Let X be a random variable and let X_ε, $\varepsilon > 0$, be defined by* (1). *If X_ε has finite expectation for some $\varepsilon > 0$, we say that X has finite expectation and define its expectation EX by*

$$EX = \lim_{\varepsilon \to 0} EX_\varepsilon.$$

Otherwise we say that X does not have finite expectation.

From the discussion preceding Theorem 1 it follows that the definition of EX can be given in terms of the distribution function of X and that if two random variables have the same distribution function, then their expectations are equal (or both not finite). Using techniques from the theory of measure and integration, we can show that Definition 2 gives the same values as do our previous definitions for the special cases when X is discrete or when X is a continuous random variable having a density. There is an analog of Theorem 1 of Chapter 4 which we state without proof. In this theorem, φ can be any function of the type considered in calculus.

Theorem 2 *Let $X_1, \ldots, X_n$ be continuous random variables having joint density f and let Z be a random variable defined in terms of $X_1, \ldots, X_n$ by $Z = \varphi(X_1, \ldots, X_n)$. Then Z has finite expectation if and only if*

$$\int_{-\infty}^{\infty} \cdots \int_{-\infty}^{\infty} |\varphi(x_1, \ldots, x_n)| f(x_1, \ldots, x_n)\, dx_1 \cdots dx_n < \infty,$$

in which case

$$EZ = \int_{-\infty}^{\infty} \cdots \int_{-\infty}^{\infty} \varphi(x_1, \ldots, x_n) f(x_1, \ldots, x_n)\, dx_1 \cdots dx_n.$$

We can show that the basic properties of expectation proven in Chapter 4 for discrete random variables are valid in general. In particular Theorems 2, 3, and 4 of Chapter 4 are valid and will be freely used.

As in the discrete case we sometimes refer to EX as the *mean* of X. The definition of moments, central moments, variance, standard deviation, covariance, and correlation given in Chapter 4 for discrete random variables depend only on the notion of expectation and extend immediately to the general case.

In general, as in the discrete case, if X has a moment of order r, then X has a moment of order k for all $k \le r$. Theorems 6 and 7 of Chapter 4 are also true in general. The reader should review these definitions and theorems in Chapter 4 before going on to the next section.

7.3. Moments of continuous random variables

Let X be a continuous random variable having density f and mean μ. If X has finite mth moment, then by Theorem 2

$$EX^m = \int_{-\infty}^{\infty} x^m f(x)\,dx$$

and

$$E(X - \mu)^m = \int_{-\infty}^{\infty} (x - \mu)^m f(x)\,dx.$$

In particular, if X has finite second moment, its variance σ^2 is given by

$$\sigma^2 = \int_{-\infty}^{\infty} (x - \mu)^2 f(x)\,dx.$$

Note that $\sigma^2 > 0$. For if $\sigma^2 = 0$, then it follows by the argument of Section 4.3 that $P(X = \mu) = 1$, which contradicts the assumption that X is a continuous random variable.

Example 4. Let X have the gamma density $\Gamma(\alpha, \lambda)$. Find the moments and the variance of X.

The mth moment of X is given by

$$EX^m = \int_0^{\infty} x^m \frac{\lambda^\alpha}{\Gamma(\alpha)} x^{\alpha-1} e^{-\lambda x}\,dx$$

$$= \frac{\lambda^\alpha}{\Gamma(\alpha)} \int_0^{\infty} x^{m+\alpha-1} e^{-\lambda x}\,dx,$$

so by Formulas (34) and (36) of Chapter 5

$$EX^m = \frac{\lambda^\alpha \Gamma(m + \alpha)}{\lambda^{m+\alpha}\Gamma(\alpha)} \tag{2}$$

$$= \frac{\alpha(\alpha + 1)\cdots(\alpha + m - 1)}{\lambda^m}.$$

The variance of X is given by

$$\sigma^2 = EX^2 - (EX)^2 = \frac{\alpha(\alpha + 1)}{\lambda^2} - \left(\frac{\alpha}{\lambda}\right)^2 = \frac{\alpha}{\lambda^2}.$$

By setting $\alpha = 1$, we see that if X has the exponential density with parameter λ, then $EX^m = m!\,\lambda^{-m}$ and X has variance λ^{-2}. For a second special case, let X have the $\chi^2(n)$ distribution which, according to Section 6.6, is the same as the $\Gamma(n/2, 1/2)$ distribution. Then

$$EX = \frac{n/2}{1/2} = n \quad \text{and} \quad \text{Var } X = \frac{n/2}{(1/2)^2} = 2n.$$

We can often take advantage of symmetry in computing moments. For example, if X has a symmetric density, if EX^m exists, and if m is an odd positive integer, then $EX^m = 0$. To see this, note that by Theorem 2 of Chapter 5, X and $-X$ have the same distribution function. Thus X^m and $(-X)^m = -X^m$ have the same distribution function and consequently the same expectation. In other words $EX^m = E(-X^m) = -EX^m$, which implies that $EX^m = 0$.

Example 5. Let X have the normal density $n(\mu, \sigma^2)$. Find the mean and central moments of X.

The random variable $X - \mu$ has the normal density $n(0, \sigma^2)$, which is a symmetric density. Thus $E(X - \mu)^m = 0$ for m an odd positive integer. In particular $E(X - \mu) = 0$, so we see that the parameter μ in the normal density $n(\mu, \sigma^2)$ is just the mean of the density. It now follows that all the odd central moments of X equal zero. To compute the even central moments we recall from Section 5.3.3 that $Y = (X - \mu)^2$ has the gamma density $\Gamma(1/2, 1/2\sigma^2)$. Since for m even $E(X - \mu)^m = EY^{m/2}$, it follows from Example 4 that

$$E(X - \mu)^m = \frac{\Gamma\left(\frac{m+1}{2}\right)}{\left(\frac{1}{2\sigma^2}\right)^{m/2}\Gamma\left(\frac{1}{2}\right)}$$

$$= \frac{\frac{1}{2}\cdot\frac{3}{2}\cdots\left(\frac{m-1}{2}\right)}{\left(\frac{1}{2\sigma^2}\right)^{m/2}}$$

$$= \sigma^m 1 \cdot 3 \cdots (m - 1).$$

By using Formulas (35) and (38) of Chapter 5, we obtain the alternative formula

$$E(X - \mu)^m = \frac{m!}{2^{m/2}\left(\frac{m}{2}\right)!}\sigma^m. \tag{3}$$

In particular σ^2 denotes the variance of X and $E(X - \mu)^4 = 3\sigma^4$.

Let X and Y be continuous random variables having joint density f, means μ_X and μ_Y, and finite second moments. Then their covariance is given by

$$E(X - \mu_X)(Y - \mu_Y) = \int_{-\infty}^{\infty}\int_{-\infty}^{\infty}(x - \mu_Y)(y - \mu_Y)f(x, y)\,dx\,dy. \tag{4}$$

Example 6. Let X and Y have the joint density f in Example 2 of Chapter 6. Find the correlation between X and Y.

According to Example 2 of Chapter 6

$$f(x, y) = \frac{\sqrt{3}}{4\pi} e^{-[(x^2 - xy + y^2)/2]}$$
$$= \frac{\sqrt{3}}{4\pi} e^{-3x^2/8} e^{-[(y-x/2)^2/2]}.$$

We saw in that example that X and Y each have the normal density $n(0, 4/3)$. Thus $\mu_X = \mu_Y = 0$ and $\text{Var } X = \text{Var } Y = 4/3$. From Equation (4) and the second expression for f, we have

$$EXY = \frac{\sqrt{3}}{2\sqrt{2\pi}} \int_{-\infty}^{\infty} xe^{-(3x^2/8)}\, dx \int_{-\infty}^{\infty} y \frac{1}{\sqrt{2\pi}} e^{-[(y-x/2)^2/2]}\, dy.$$

Now

$$\int_{-\infty}^{\infty} y \frac{1}{\sqrt{2\pi}} e^{-[(y-x/2)^2/2]}\, dy = \int_{-\infty}^{\infty} \left(u + \frac{x}{2}\right) \frac{1}{\sqrt{2\pi}} e^{-(u^2/2)}\, du = \frac{x}{2},$$

and hence

$$EXY = \frac{1}{2\left(\frac{2}{\sqrt{3}}\right)\sqrt{2\pi}} \int_{-\infty}^{\infty} x^2 e^{-(3x^2/8)}\, dx$$
$$= 1/2 \int_{-\infty}^{\infty} x^2 n(x; 0, 4/3)\, dx$$
$$= 1/2 \cdot 4/3 = 2/3.$$

The correlation ρ between X and Y is given by

$$\rho = \frac{EXY}{\sqrt{\text{Var } X}\sqrt{\text{Var } Y}} = \frac{2/3}{\sqrt{4/3}\sqrt{4/3}} = \frac{1}{2}.$$

Example 7. Let $U_1, \ldots, U_n$ be independent random variables each uniformly distributed over $(0, 1)$ and set

$$X = \min(U_1, \ldots, U_n)$$

and

$$Y = \max(U_1, \ldots, U_n).$$

Find the moments of X and Y and the correlation between X and Y.

These random variables were studied in Section 6.5 (where they were denoted by X_1 and X_n). Specializing the results of that section to U_i's which are uniformly distributed, we find that X and Y have a joint density f given by

$$(5) \qquad f(x, y) = \begin{cases} n(n-1)(y-x)^{n-2}, & 0 \le x \le y \le 1, \\ 0, & \text{elsewhere.} \end{cases}$$

Those readers who have skipped Section 6.5 can think of the present problem as that of finding the moments of, and the correlation between two random variables X and Y whose joint density is given by (5).

The mth moment of X is given by

$$\begin{aligned} EX^m &= n(n-1)\int_0^1 x^m\,dx \int_x^1 (y-x)^{n-2}\,dy \\ &= n(n-1)\int_0^1 x^m\,dx \left.\frac{(y-x)^{n-1}}{n-1}\right|_{y=x}^{y=1} \\ &= n\int_0^1 x^m(1-x)^{n-1}\,dx. \end{aligned}$$

The definite integral appearing in this expression is a Beta integral and was evaluated in Formula (19) of Chapter 6. From this formula we find that

$$EX^m = \frac{n\Gamma(m+1)\Gamma(n)}{\Gamma(m+n+1)} = \frac{m!\,n!}{(m+n)!}.$$

In particular, $EX = 1/(n+1)$ and $EX^2 = 2/(n+1)(n+2)$. It follows that

$$\operatorname{Var} X = (EX^2) - (EX)^2 = \frac{n}{(n+1)^2(n+2)}.$$

The mth moment of Y is given by

$$\begin{aligned} EY^m &= n(n-1)\int_0^1 y^m\,dy \int_0^y (y-x)^{n-2}\,dx \\ &= n(n-1)\int_0^1 y^m\,dy \left.\frac{(y-x)^{n-1}(-1)}{n-1}\right|_{x=0}^{x=y} \\ &= n\int_0^1 y^{m+n-1}\,dy \\ &= \frac{n}{m+n}. \end{aligned}$$

Thus $EY = n/(n+1)$ and

$$\operatorname{Var} Y = \frac{n}{n+2} - \left(\frac{n}{n+1}\right)^2 = \frac{n}{(n+1)^2(n+2)}.$$

Alternatively, these quantities could be computed from the marginal densities of X and Y.

To find the covariance of X and Y we start with

$$EXY = n(n-1)\int_0^1 y\,dy \int_0^y x(y-x)^{n-2}\,dx.$$

Since

$$x(y - x)^{n-2} = y(y - x)^{n-2} - (y - x)^{n-1},$$

we find that

$$\begin{aligned} EXY &= n(n-1)\int_0^1 y^2\,dy \int_0^y (y-x)^{n-2}\,dx \\ &\quad - n(n-1)\int_0^1 y\,dy \int_0^y (y-x)^{n-1}\,dx \\ &= n(n-1)\int_0^1 y^2\,dy \left.\frac{(y-x)^{n-1}(-1)}{n-1}\right|_{x=0}^{x=y} \\ &\quad - n(n-1)\int_0^1 y\,dy \left.\frac{(y-x)^{n}(-1)}{n}\right|_{x=0}^{x=y} \\ &= n\int_0^1 y^{n+1}\,dy - (n-1)\int_0^1 y^{n+1}\,dy \\ &= \frac{1}{n+2}. \end{aligned}$$

Consequently

$$\begin{aligned} \operatorname{Cov}(X, Y) &= EXY - EXEY \\ &= \frac{1}{n+2} - \frac{n}{(n+1)^2} \\ &= \frac{1}{(n+1)^2(n+2)}. \end{aligned}$$

Finally we obtain for the correlation between X and Y,

$$\begin{aligned} \rho &= \frac{\operatorname{Cov}(X, Y)}{\sqrt{\operatorname{Var} X \operatorname{Var} Y}} \\ &= \frac{1}{(n+1)^2(n+2)} \Big/ \frac{n}{(n+1)^2(n+2)} \\ &= \frac{1}{n}. \end{aligned}$$

7.4. Conditional expectation

Let X and Y be continuous random variables having joint density f and suppose that Y has finite expectation. In Section 6.3 we defined the conditional density of Y given $X = x$ by

$$f_{Y|X}(y \mid x) = \begin{cases} \dfrac{f(x, y)}{f_X(x)}, & 0 < f_X(x) < \infty, \\ 0, & \text{elsewhere.} \end{cases}$$

For each x such that $0 < f_X(x) < \infty$ the function $f_{Y|X}(y \mid x)$, $-\infty < y < \infty$, is a density function according to Definition 5 of Chapter 5. Thus we can talk about the various moments of this density. Its mean is called the *conditional expectation* of Y given $X = x$ and is denoted by $E[Y \mid X = x]$ or $E[Y \mid x]$. Thus

$$E[Y \mid X = x] = \int_{-\infty}^{\infty} yf(y \mid x)\, dy \tag{6}$$

$$= \frac{\int_{-\infty}^{\infty} yf(x, y)\, dy}{f_X(x)}$$

when $0 < f_X(x) < \infty$. We define $E[Y \mid X = x] = 0$ elsewhere. In statistics the function m defined by $m(x) = E[Y \mid X = x]$ is called the *regression function* of Y on X.

Conditional expectations arise in statistical problems involving prediction and Bayesian estimation, as we will see in Volume II. They are also important, from a more general viewpoint, in advanced probability theory. We will confine ourselves to some elementary illustrations of conditional expectations. The general theory is quite sophisticated and will not be needed in this book.

Example 8. Let X and Y have the joint density f in Example 2 of Chapter 6. Find the conditional expectation of Y given $X = x$.

In Example 7 of Chapter 6 we found that the conditional density of Y given $X = x$ is the normal density $n(x/2, 1)$ which we now know has mean $x/2$. Thus

$$E[Y \mid X = x] = \frac{x}{2}.$$

In this example the conditional variance of Y given $X = x$ is the constant 1.

Example 9. Let X and Y be continuous random variables having joint density f given by (5). In the previous section we computed various moments involving X and Y. Here we will compute the conditional density and conditional expectation of Y given $X = x$.

The marginal density of X is given by

$$f_X(x) = n(n-1)\int_x^1 (y - x)^{n-2}\, dy$$

$$= n(1 - x)^{n-1}, \qquad 0 \le x \le 1,$$

and $f_X(x) = 0$ elsewhere. Thus for $0 \le x < 1$,

$$f(y \mid x) = \begin{cases} \dfrac{(n-1)(y-x)^{n-2}}{(1-x)^{n-1}}, & x \le y < 1, \\ 0, & \text{elsewhere.} \end{cases}$$

Consequently, for $0 \le x < 1$,

$$
\begin{aligned}
E[Y \mid X = x] &= \int_{-\infty}^{\infty} y f(y \mid x)\, dy \\
&= (n-1)(1-x)^{1-n} \int_x^1 y(y-x)^{n-2}\, dy \\
&= (n-1)(1-x)^{1-n} \int_x^1 [(y-x)^{n-1} + x(y-x)^{n-2}]\, dy \\
&= (n-1)(1-x)^{1-n} \left[\frac{(1-x)^n}{n} + \frac{x(1-x)^{n-1}}{n-1}\right] \\
&= \frac{(n-1)(1-x)}{n} + x \\
&= \frac{n-1+x}{n}.
\end{aligned}
$$

It is sometimes convenient to calculate the expectation of Y according to the formula

$$
EY = \int_{-\infty}^{\infty} E[Y \mid X = x] f_X(x)\, dx. \tag{7}
$$

This formula follows immediately from (6). For

$$
\begin{aligned}
\int_{-\infty}^{\infty} E[Y \mid X = x] f_X(x)\, dx &= \int_{-\infty}^{\infty} dx \int_{-\infty}^{\infty} y f(x, y)\, dy \\
&= \int_{-\infty}^{\infty} \int_{-\infty}^{\infty} y f(x, y)\, dx\, dy \\
&= EY.
\end{aligned}
$$

Applying this formula to Example 9, we get

$$
\begin{aligned}
EY &= \int_0^1 \left(\frac{n-1+x}{n}\right) n(1-x)^{n-1}\, dx \\
&= n \int_0^1 (1-x)^{n-1}\, dx - \int_0^1 (1-x)^n\, dx \\
&= 1 - \frac{1}{n+1} = \frac{n}{n+1},
\end{aligned}
$$

which agrees with the answer found in Example 7.

Naturally, conditional expectations can be defined similarly for discrete random variables. Some exercises involving this were given in Chapter 4.

7.5. The Central Limit Theorem

Throughout this section $X_1, X_2, \ldots$ will denote independent, identically distributed random variables having mean μ and finite nonzero variance

σ^2. We will be interested in studying the distribution of $S_n = X_1 + \cdots + X_n$. First of all we note that S_n has mean $n\mu$ and variance $n\sigma^2$.

Suppose next that X_1 has density f. Then, for all $n \geq 1$, S_n will have a density f_{S_n}. Now $f_{S_1} = f$, and the other densities can be calculated successively by using formulas obtained in Chapters 3 and 6 for the density of the sum of two independent random variables. We have that

$$f_{S_n}(x) = \sum_y f_{S_{n-1}}(y)f(x - y)$$

or

$$f_{S_n}(x) = \int_{-\infty}^{\infty} f_{S_{n-1}}(y)f(x - y)\, dy$$

according as X_1 is a discrete or a continuous random variable. For certain choices of f (e.g., binomial, negative binomial, Poisson, normal, and gamma), we can find simple formulas for f_{S_n}. In general, however, we have to resort to numerical methods.

One of the most important and most remarkable results of probability theory is that for large values of n the distribution of S_n depends on the distribution of X_1 essentially only through μ and σ^2. Such a result is more easily discussed in terms of the normalized random variable

$$S_n^* = \frac{S_n - ES_n}{\sqrt{\operatorname{Var} S_n}} = \frac{S_n - n\mu}{\sigma\sqrt{n}},$$

which has mean 0 and variance 1.

To get some idea of how the distribution function of S_n^* behaves as $n \to \infty$, let us first consider a special case in which this distribution function can be found easily and exactly. Suppose, then, that X_1 is normally distributed with mean μ and variance σ^2. Then, by results in Chapter 6, S_n^* is normally distributed with mean 0 and variance 1 or, in other words, S_n^* has the standard normal distribution function Φ.

Suppose next that X_1 takes values 1 and 0 with respective probabilities p and $1 - p$. Then as we saw in Chapter 3, S_n has a binomial distribution with parameters n and p; that is,

$$P(S_n = k) = \binom{n}{k} p^k(1 - p)^{n-k}.$$

It was discovered by DeMoivre (1667–1754) and Laplace (1749–1827) that in this case the distribution function of S_n^* approaches Φ, the standard normal distribution function, as $n \to \infty$.

In more recent times there have been many extensions of the DeMoivre–Laplace limit theorem, all known as "central limit theorems." The simplest and best known of these results was proven by Lindeberg in 1922:

Theorem 3 Central Limit Theorem. *Let $X_1, X_2, \ldots$ be independent, identically distributed random variables having mean μ and finite nonzero variance σ^2. Set $S_n = X_1 + \cdots + X_n$. Then*

$$(8) \qquad \lim_{n\to\infty} P\left(\frac{S_n - n\mu}{\sigma\sqrt{n}} \le x\right) = \Phi(x), \qquad -\infty < x < \infty.$$

The generality of this theorem is remarkable. The random variable X_1 can be discrete, continuous, or neither of these. Moreover, the conclusion holds even if no moments of X_1 exist beyond the second. Another very surprising part of the theorem is that the limiting distribution function of S_n^* is independent of the specific distribution of X_1 (provided, of course, that the hypotheses of the theorem are satisfied). We should not be surprised, however, that Φ is that limiting distribution. For we have seen that this is true if X_1 has either a normal or a binomial distribution.

The proof of the Central Limit Theorem will be postponed to Chapter 8, since it requires advanced techniques yet to be discussed which involve characteristic functions. It is possible to give an elementary but somewhat laborious proof of the DeMoivre–Laplace limit theorem, the special case of the Central Limit Theorem when X_1 is binomially distributed. There are elementary ways to make the Central Limit Theorem plausible, but they are not proofs. One such way is to show that, for any positive integer m, if X_1 has finite mth moment, then

$$\lim_{n\to\infty} E\left(\frac{S_n - n\mu}{\sigma\sqrt{n}}\right)^m$$

exists and equals the mth moment of the standard normal distribution.

At this stage it is more profitable to understand what the Central Limit Theorem means and how it can be used in typical applications.

Example 10. Let $X_1, X_2, \ldots$ be independent random variables each having a Poisson distribution with parameter λ. Then by results in Chapter 4, $\mu = \sigma^2 = \lambda$ and S_n has a Poisson distribution with parameter $n\lambda$. The Central Limit Theorem implies that

$$\lim_{n\to\infty} P\left(\frac{S_n - n\lambda}{\sqrt{n\lambda}} \le x\right) = \Phi(x), \qquad -\infty < x < \infty.$$

One can extend the result of this example and show that if X_t is a random variable having a Poisson distribution with parameter $\lambda = t$, then

$$(9) \qquad \lim_{t\to\infty} P\left(\frac{X_t - EX_t}{\sqrt{\operatorname{Var} X_t}} \le x\right) = \Phi(x), \qquad -\infty < x < \infty.$$

Equation (9) also holds if X_t is a random variable having the gamma distribution $\Gamma(t, \lambda)$ for fixed λ, or the negative binomial distribution with parameters $\alpha = t$ and p fixed.

7.5.1. Normal approximations. The Central Limit Theorem strongly suggests that for large n we should make the approximation

$$P\left(\frac{S_n - n\mu}{\sigma\sqrt{n}} \le x\right) \approx \Phi(x), \qquad -\infty < x < \infty,$$

or equivalently

$$P(S_n \le x) \approx \Phi\left(\frac{x - n\mu}{\sigma\sqrt{n}}\right) = \Phi\left(\frac{x - ES_n}{\sqrt{\operatorname{Var} S_n}}\right), \qquad -\infty < x < \infty. \tag{10}$$

We will refer to (10) as a *normal approximation* formula. According to this formula we approximate the distribution function of S_n by the normal distribution function having the same mean and variance. One difficulty in applying the normal approximation formula is in deciding how large n must be for (10) to be valid to some desired degree of accuracy. Various numerical studies have indicated that in typical practical applications $n = 25$ is sufficiently large for (10) to be useful.

As an example where normal approximation is applicable, let X_1, $X_2, \ldots$ be independent random variables each having an exponential density with parameter $\lambda = 1$. Then (10) becomes

$$P(S_n \le x) \approx \Phi\left(\frac{x - n}{\sqrt{n}}\right), \qquad -\infty < x < \infty. \tag{11}$$

Graphs showing the accuracy of this approximation are given in Figure 1 for $n = 10$.

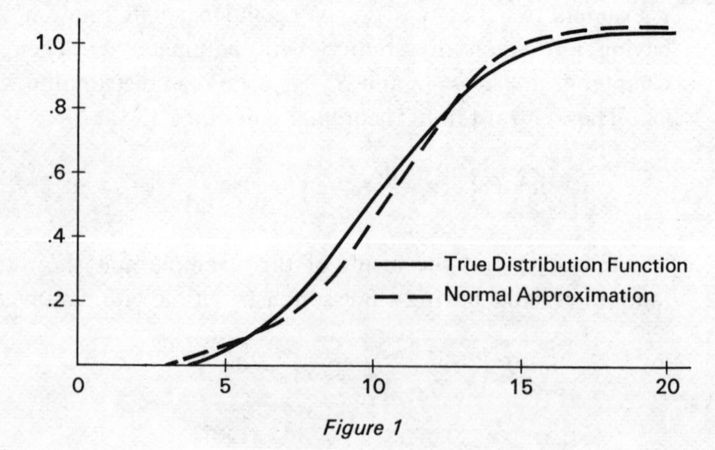

Figure 1

Example 11. Suppose the length of life of a certain kind of light bulb, after it is installed, is exponentially distributed with a mean length of 10 days. As soon as one light bulb burns out, a similar one is installed in its place. Find the probability that more than 50 bulbs will be required during a one-year period.

In solving this problem we let X_n denote the length of life of the nth light bulb that is installed. We assume that $X_1, X_2, \ldots$ are independent random variables each having an exponential distribution with mean 10 or parameter $\lambda = 1/10$. Then $S_n = X_1 + \cdots + X_n$ denotes the time when the nth bulb burns out. We want to find $P(S_{50} < 365)$. Now S_{50} has mean $50\lambda^{-1} = 500$ and variance $50\lambda^{-2} = 5000$. Thus by the normal approximation formula (10)

$$P(S_{50} < 365) \approx \Phi\left(\frac{365 - 500}{\sqrt{5000}}\right)$$

$$= \Phi(-1.91) = .028.$$

It is therefore very unlikely that more than 50 bulbs will be needed.

Suppose that S_n is a continuous random variable having density f_{S_n}. If we differentiate the terms in (10) we obtain

$$f_{S_n}(x) \approx \frac{1}{\sigma\sqrt{n}}\, \varphi\left(\frac{x - n\mu}{\sigma\sqrt{n}}\right), \qquad -\infty < x < \infty. \tag{12}$$

Though the derivation of (12) is far from a proof, (12) is actually a good approximation for n large (under the further mild restriction that, for some n, f_{S_n} is a bounded function).

As an example of this approximation let X_1 be exponentially distributed with parameter $\lambda = 1$, so that (11) is applicable. Then (12) becomes

$$f_{S_n}(x) \approx \frac{1}{\sqrt{n}}\, \varphi\left(\frac{x - n}{\sqrt{n}}\right), \qquad -\infty < n < \infty. \tag{13}$$

Graphs showing the accuracy of this approximation are given in Figure 2 for $n = 10$.

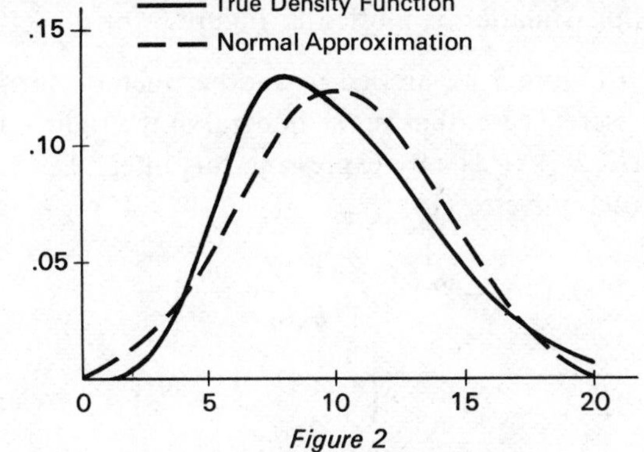

Figure 2

Forms of the Central Limit Theorem involving densities instead of distribution functions are known as "local" central limit theorems. They are also important, especially in the advanced theory of probability.

There is an approximation similar to (12) for discrete random variables. Naturally, a precise statement of such an approximation depends on the nature of the possible values of S_n, that is, those values of x such that $f_{S_n}(x) = P(S_n = x) > 0$. For simplicity we make the following two assumptions:

(i) If x is a possible value of X_1, then x is an integer;

(ii) if a is a possible value of X_1, then the greatest common divisor of the set

$$\{x - a \mid x \text{ is a possible value of } X_1\}$$

is one.

We exclude, for example, a random variable X_1 such that $P(X_1 = 1) = P(X_1 = 3) = 1/2$, for then the greatest common divisor of the indicated set is 2. Under assumptions (i) and (ii), the approximation

$$(14) \qquad f_{S_n}(x) \approx \frac{1}{\sigma\sqrt{n}}\,\varphi\left(\frac{x - n\mu}{\sigma\sqrt{n}}\right), \qquad x \text{ an integer,}$$

is valid for large n.

Example 12. Let X_1 be the binomial random variable taking on values 1 and 0 with probabilities p and $1 - p$ respectively. Then (i) and (ii) hold and (14) is applicable with $\mu = p$ and $\sigma^2 = p(1 - p)$. Since S_n has the binomial distribution with parameters n and p, we have the approximation

$$(15) \qquad f_{S_n}(x) = \binom{n}{x} p^x(1 - p)^{n-x}$$

$$\approx \frac{1}{\sqrt{np(1 - p)}}\,\varphi\left(\frac{x - np}{\sqrt{np(1 - p)}}\right), \qquad x \text{ an integer.}$$

This approximation is plotted in Figure 3 for $n = 10$ and $p = .3$.

From Figure 3 we are led to another method for approximating $f_{S_n}(x)$ in the discrete case, that is, the integral of the right side of (14) over the set $[x - 1/2, x + 1/2]$. By expressing this integral in terms of Φ we obtain as an alternative to (14)

$$(16) \qquad f_{S_n}(x) \approx \Phi\left(\frac{x + (1/2) - n\mu}{\sigma\sqrt{n}}\right)$$

$$- \Phi\left(\frac{x - (1/2) - n\mu}{\sigma\sqrt{n}}\right), \qquad x \text{ an integer.}$$

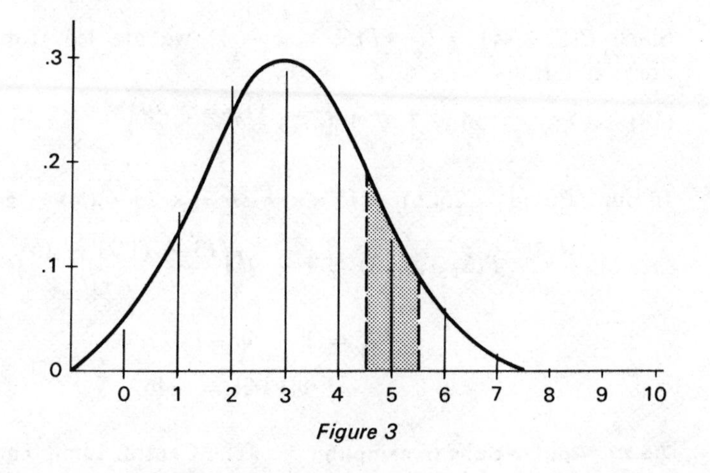

Figure 3

The area of the shaded region of Figure 3 is an approximation to $P(S_n = 5)$.

Finally, if we sum (16) on the set $\{\ldots, x - 2, x - 1, x\}$ we are led to the approximation

$$(17) \qquad P(S_n \leq x) \approx \Phi\left(\frac{x + (1/2) - n\mu}{\sigma\sqrt{n}}\right), \qquad x \text{ an integer.}$$

When S_n is discrete and conditions (i) and (ii) hold, then (17) is usually more accurate than is the original normal approximation Formula (10). In Figure 4 we compare the approximations in Formulas (10) and (17) when S_n has the binomial distribution with parameters $n = 10$ and $p = .3$.

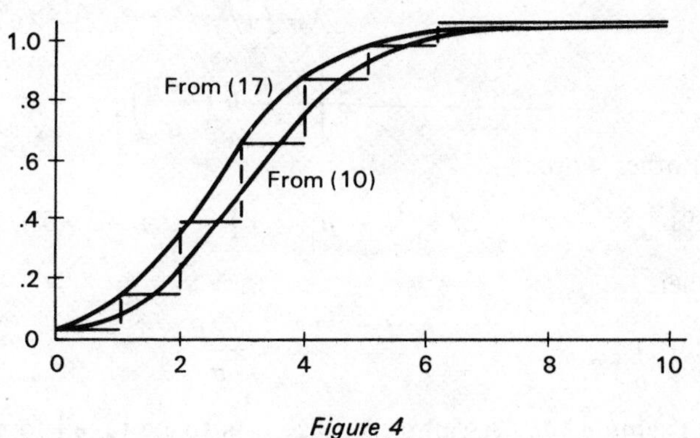

Figure 4

Example 13. A certain basketball player knows that on the average he will make 60 percent of his freethrow attempts. What is the probability that in 25 attempts he will be successful more than one-half of the time?

We will interpret the problem as implying that the number S_n of successes in n attempts is binomially distributed with parameters n and $p = .6$.

Since $P(S_n \geq x) = 1 - P(S_n \leq x - 1)$ we are led from (17) to the approximation

(18) $$P(S_n \geq x) \approx 1 - \Phi\left(\frac{x - (1/2) - n\mu}{\sigma\sqrt{n}}\right), \qquad x \text{ an integer.}$$

In our case $n\mu = 25(.6) = 15$ and $\sigma\sqrt{n} = \sqrt{25(.6)(.4)} = 5\sqrt{.24}$. Thus

$$\begin{aligned} P(S_{25} \geq 13) &\approx 1 - \Phi\left(\frac{13 - (1/2) - 15}{5\sqrt{.24}}\right) \\ &= 1 - \Phi(-1.02) \\ &= \Phi(1.02) = .846. \end{aligned}$$

7.5.2. Applications to sampling. The Central Limit Theorem and the corresponding normal approximation formulas can be regarded as refinements of the Weak Law of Large Numbers discussed in Chapter 4. We recall that this law states that for large n, S_n/n should be close to μ with probability close to 1. The weak law itself, however, provides no information on how accurate such an estimate should be. As we saw in Chapter 4, Chebyshev's Inequality sheds some light on this question.

The normal approximation formula (10) is also useful in this context. For $c > 0$

$$\begin{aligned} P\left(\left|\frac{S_n}{n} - \mu\right| \geq c\right) &= P(S_n \leq n\mu - nc) + P(S_n \geq n\mu + nc) \\ &\approx \Phi\left(\frac{-nc}{\sigma\sqrt{n}}\right) + 1 - \Phi\left(\frac{nc}{\sigma\sqrt{n}}\right) \\ &= 2\left[1 - \Phi\left(\frac{c\sqrt{n}}{\sigma}\right)\right]. \end{aligned}$$

In other words

(19) $$P\left(\left|\frac{S_n}{n} - \mu\right| \geq c\right) \approx 2(1 - \Phi(\delta)),$$

where

(20) $$\delta = \frac{c\sqrt{n}}{\sigma}.$$

Example 14. A sample of size n is to be taken to determine the percentage of the population planning to vote for the incumbent in an election. Let $X_k = 1$ if the kth person sampled plans to vote for the incumbent and $X_k = 0$ otherwise. We assume that $X_1, \ldots, X_n$ are independent, identically distributed random variables such that $P(X_1 = 1) = p$ and $P(X_1 = 0) = 1 - p$. Then $\mu = p$ and $\sigma^2 = p(1 - p)$. We will also assume that p is close enough to .5 so that $\sigma = \sqrt{p(1 - p)}$ can be approximated satisfactorily by $\sigma \approx 1/2$ (note that σ has a maximum of 1/2 at $p = .5$, and that, as p ranges over $.3 \le p \le .7$, σ stays above .458 which is close to 1/2). The random variable S_n/n denotes the fraction of people sampled that plan to vote for the incumbent and can be used to estimate the true but unknown probability p. We will use normal approximations to solve the following three problems:

(i) Suppose $n = 900$. Find the probability that

$$\left|\frac{S_n}{n} - p\right| \ge .025.$$

(ii) Suppose $n = 900$. Find c such that

$$P\left(\left|\frac{S_n}{n} - p\right| \ge c\right) = .01.$$

(iii) Find n such that

$$P\left(\left|\frac{S_n}{n} - p\right| \ge .025\right) = .01.$$

Solution to (i). By (20)

$$\delta = \frac{(.025)\sqrt{900}}{.5} = 1.5,$$

so by (19)

$$P\left(\left|\frac{S_n}{n} - p\right| \ge .025\right) \approx 2(1 - \Phi(1.5))$$

$$= 2(.067) = .134.$$

Solution to (ii). We first choose δ so that $2(1 - \Phi(\delta)) = .01$ or $\Phi(\delta) = .995$. Inspection of Table I shows that $\delta = 2.58$. Solving (20) for c we get

$$c = \frac{\delta\sigma}{\sqrt{n}} = \frac{(2.58)(.5)}{\sqrt{900}} = .043.$$

Solution to (iii). As in (ii) we have $\delta = 2.58$. Solving (20) for n we find

$$n = \frac{\delta^2\sigma^2}{c^2} = \frac{(2.58)^2(.25)}{(.025)^2} = 2663.$$

It is worthwhile to compare the results of (ii) and (iii). In both cases

$$P\left(\left|\frac{S_n}{n} - p\right| \ge c\right) \approx .01.$$

In (ii), $c = .043$ and $n = 900$, while in (iii), $c = .025$ and $n = 2663$. In going from (ii) to (iii), in order to decrease c by the factor 43/25 we are forced to increase n by the square of that factor. This is true generally whenever we want to keep

$$P\left(\left|\frac{S_n}{n} - \mu\right| \geq c\right)$$

constant. For then δ is determined by (19) and, from (20), n is related to c by $n = \delta^2\sigma^2/c^2$. In the same context, if we increase n by some factor, we decrease c only by the square root of that factor.

Exercises

1 Let X have a Beta density with parameters α_1 and α_2. Find EX.

2 Let X and Y be independent random variables having respective gamma densities $\Gamma(\alpha_1, \lambda)$ and $\Gamma(\alpha_2, \lambda)$. Set $Z = Y/X$. For which values of α_1 and α_2 will Z have finite expectation? Find EZ when it exists. *Hint:* See Theorem 3 of Chapter 6 and related discussion.

3 Let X have the normal density $n(0, \sigma^2)$. Find $E|X|$. *Hint:* Use the result of Exercise 31 of Chapter 5.

4 Let X have an exponential density with parameter λ and let X_ε be defined in terms of X and $\varepsilon > 0$ by (1). What is the distribution of $X_\varepsilon/\varepsilon$? Find EX_ε and evaluate its limit as $\varepsilon \to 0$.

5 Let X have a Beta density with parameters α_1 and α_2. Find the moments and the variance of X.

6 Let X have a χ^2 distribution with n degrees of freedom. Find the mean of $Y = \sqrt{X}$.

7 Let X be the random variable as in Example 7. Find EX^m from the marginal density f_X.

8 Let Z be as in Exercise 2. Find the variance of Z.

9 Let U_1 and U_2 be independent random variables each having an exponential density with parameter λ, and set $Y = \max(U_1, U_2)$. Find the mean and variance of Y (see Section 6.5).

10 Let X be the random variable from Example 1 of Chapter 5. Find the mean and variance of X.

11 Let X be the random variable from Example 1 of Chapter 6. Find the mean and variance of X. *Hint:* Reduce EX^2 to a Beta integral.

12 Find the mean and variance of the random variable Z from Exercise 17 of Chapter 6.

13 Find the mean and variance of the random variable Y from Exercise 28 of Chapter 6.

14 Let X be the sine of an angle in radians chosen uniformly from $(-\pi/2, \pi/2)$. Find the mean and variance of X.

15 Let X have the normal density $n(0, \sigma^2)$. Find the mean and variance of each of the following random variables:
(a) $|X|$;
(b) X^2;
(c) e^{tX}.

16 Let X have the gamma density $\Gamma(\alpha, \lambda)$. For which real t does e^{tX} have finite expectation? Find Ee^{tX} for these values of t.

17 Let X have the gamma density $\Gamma(\alpha, \lambda)$. For which real numbers r does X^r have finite expectation? Find EX^r for these values of r.

18 Let X be a nonnegative continuous random variable having density f and distribution function F. Show that X has finite expectation if and only if

$$\int_0^\infty (1 - F(x))\, dx < \infty$$

and then

$$EX = \int_0^\infty (1 - F(x))\, dx.$$

Hint: See the proof of Theorem 5 of Chapter 4.

19 Let X_k be the kth order statistic in a sample from random variables $U_1, \ldots, U_n$ which are independent and uniformly distributed over $(0, 1)$. Find the mean and variance of X_k.

20 Let X and Y be as in Example 7 and let $R = Y - X$. Find the mean and variance of R. *Hint:* Use Equation (16) of Chapter 4.

21 Let X and Y have density f as in Exercise 9 of Chapter 6. Find the correlation between X and Y.

22 Let X and Y be independent random variables such that X has the normal density $n(\mu, \sigma^2)$ and Y has the gamma density $\Gamma(\alpha, \lambda)$. Find the mean and variance of the random variable $Z = XY$.

23 Let X and Y be random variables having mean 0, variance 1, and correlation ρ. Show that $X - \rho Y$ and Y are uncorrelated, and that $X - \rho Y$ has mean 0 and variance $1 - \rho^2$.

24 Let X, Y, and Z be random variables having mean 0 and unit variance. Let ρ_1 be the correlation between X and Y, ρ_2 the correlation between Y and Z, and ρ_3 the correlation between X and Z. Show that

$$\rho_3 \geq \rho_1\rho_2 - \sqrt{1 - \rho_1^2}\,\sqrt{1 - \rho_2^2}.$$

Hint: Write

$$XZ = [\rho_1 Y + (X - \rho_1 Y)][\rho_2 Y + (Z - \rho_2 Y)],$$

and use the previous problem and Schwarz's inequality.

25 Let X, Y, and Z be as in the previous problem. Suppose $\rho_1 \geq .9$ and $\rho_2 \geq .8$. What can be said about ρ_3?

26 Let X and Y have a density f that is uniform over the interior of the triangle with vertices at (0, 0), (2, 0), and (1, 2). Find the conditional expectation of Y given X.

27 Let X and Y be independent random variables having respective gamma densities $\Gamma(\alpha_1, \lambda)$ and $\Gamma(\alpha_2, \lambda)$, and set $Z = X + Y$. Find the conditional expectation of X given Z.

28 Let Π and Y be random variables as in Exercise 26 of Chapter 6. Find the conditional expectation of Π given Y.

29 Let X and Y be continuous random variables having a joint density. Suppose that Y and $\varphi(X)Y$ have finite expectation. Show that

$$E\varphi(X)Y = \int_{-\infty}^{\infty} \varphi(x)E[Y \mid X = x]f_X(x)\,dx.$$

30 Let X and Y be continuous random variables having a joint density, and let $\operatorname{Var}[Y \mid X = x]$ denote the variance of the conditional density of Y given $X = x$. Show that if $E[Y \mid X = x] = \mu$ independently of X, then $EY = \mu$ and

$$\operatorname{Var} Y = \int_{-\infty}^{\infty} \operatorname{Var}[Y \mid X = x]f_X(x)\,dx.$$

31 Let $X_1, X_2, \ldots$ be independent, identically distributed random variables having mean 0 and finite nonzero variance σ^2 and set $S_n = X_1 + \cdots + X_n$. Show that if X_1 has finite third moment, then $ES_n^3 = nEX_1^3$ and

$$\lim_{n\to\infty} E\left(\frac{S_n}{\sigma\sqrt{n}}\right)^3 = 0,$$

which is the third moment of the standard normal distribution.

32 Let $X_1, \ldots, X_n$, and S_n be as in Exercise 31. Show that if X_1 has finite fourth moment, then

$$ES_n^4 = nEX_1^4 + 3n(n-1)\sigma^4$$

and

$$\lim_{n\to\infty} E\left(\frac{S_n}{\sigma\sqrt{n}}\right)^4 = 3,$$

which is the fourth moment of the standard normal distribution. *Hint:* The term $3n(n-1)$ comes from the expression

$$\binom{n}{2}\frac{4!}{2!\,2!}.$$

33 Let X have the gamma density $\Gamma(\alpha, \lambda)$. Find the normal approximation for $P(X \le x)$.

34 Let $X_1, X_2, \ldots$ be independent, normally distributed random variables having mean 0 and variance σ^2.

(a) What is the mean and variance of the random variable X_1^2?
(b) How should $P(X_1^2 + \cdots + X_n^2 \le x)$ be approximated in terms of Φ?

35 Let $X_1, X_2, \ldots$ be independent normally distributed random variables having mean 0 and variance 1 (see previous exercise).
(a) Find $P(X_1^2 + \cdots + X_{100}^2 \le 120)$.
(b) Find $P(80 \le X_1^2 + \cdots + X_{100}^2 \le 120)$.
(c) Find c such that $P(X_1^2 + \cdots + X_{100}^2 \le 100 + c) = .95$.
(d) Find c such that $P(100 - c \le X_1^2 + \cdots + X_{100}^2 \le 100 + c) = .95$.

36 A runner attempts to pace off 100 meters for an informal race. His paces are independently distributed with mean $\mu = .97$ meters and standard deviation $\sigma = .1$ meter. Find the probability that his 100 paces will differ from 100 meters by no more than 5 meters.

37 Twenty numbers are rounded off to the nearest integer and then added. Assume the individual round-off errors are independent and uniformly distributed over $(-1/2, 1/2)$. Find the probability that the given sum will differ from the sum of the original twenty numbers by more than 3.

38 A fair coin is tossed until 100 heads appear. Find the probability that at least 226 tosses will be necessary.

39 In the preceding problem find the probability that exactly 226 tosses will be needed.

40 Let X have a Poisson distribution with parameter λ.
(a) How should $f_X(x)$ be approximated in terms of the standard normal density φ?
(b) How should $f_X(x)$ be approximated in terms of the standard normal distribution function Φ?

41 Let S_n have a binomial distribution with parameters n and $p = 1/2$. How does $P(S_{2n} = n)$ behave for n large? *Hint:* Use approximation (15).

42 Players A and B make a series of \$1 bets which each player has probability 1/2 of winning. Let S_n be the net amount won by player A after n games. How does $P(S_{2n} = 0)$ behave for n large? *Hint:* See previous problem. Why isn't approximation (15) directly applicable in this case?

43 Candidates A and B are running for office and 55% of the electorate favor candidate B. What is the probability that in a sample of size 100 at least one-half of those sampled will favor candidate A?

44 A polling organization samples 1200 voters to estimate the proportion planning to vote for candidate A in a certain election. How large would the true proportion p have to be for candidate A to be 95% sure that the majority of those sampled will vote for him?

45 Suppose candidate A from the preceding problem insisted that the sample size be increased to a number n such that if 51% of all voters favored him he could be 95% sure of getting a majority of the votes sampled. About how large would n have to be?

46 Solve Exercise 27 of Chapter 4 by using normal approximation.

8 Moment Generating Functions and Characteristic Functions

Some of the most important tools in probability theory are borrowed from other branches of mathematics. In this chapter we discuss two such closely related tools. We begin with moment generating functions and then treat characteristic functions. The latter are somewhat more difficult to understand at an elementary level because they require the use of complex numbers. It is worthwhile, however, to overcome this obstacle, for a knowledge of the properties of characteristic functions will enable us to prove both the Weak Law of Large Numbers and the Central Limit Theorem (Section 8.4).

8.1. Moment generating functions

The *moment generating function* $M_X(t)$ of a random variable X is defined by

$$M_X(t) = Ee^{tX}.$$

The domain of M_X is all real numbers t such that e^{tX} has finite expectation.

Example 1. Let X be normally distributed with mean μ and variance σ^2. Then

$$\begin{aligned} M_X(t) = Ee^{tX} &= \int_{-\infty}^{\infty} e^{tx} \frac{1}{\sigma\sqrt{2\pi}} e^{-[(x-\mu)^2/2\sigma^2]}\, dx \\ &= \int_{-\infty}^{\infty} e^{t(y+\mu)} \frac{1}{\sigma\sqrt{2\pi}} e^{-y^2/2\sigma^2}\, dy \\ &= e^{\mu t} \int_{-\infty}^{\infty} \frac{1}{\sigma\sqrt{2\pi}} e^{ty-(y^2/2\sigma^2)}\, dy. \end{aligned}$$

Now

$$ty - \frac{y^2}{2\sigma^2} = -\frac{(y-\sigma^2 t)^2}{2\sigma^2} + \frac{\sigma^2 t^2}{2}.$$

Consequently

$$M_X(t) = e^{\mu t} e^{\sigma^2 t^2/2} \int_{-\infty}^{\infty} \frac{1}{\sigma\sqrt{2\pi}} e^{-[(y-\sigma^2 t)^2/2\sigma^2]}\, dt.$$

Since the last integral represents the integral of the normal density $n(\sigma^2 t, \sigma^2)$, its value is one and therefore

$$M_X(t) = e^{\mu t} e^{\sigma^2 t^2/2}, \qquad -\infty < t < \infty. \tag{1}$$

Example 2. Let X have the gamma density with parameters α and λ. Then

$$\begin{aligned} M_X(t) &= \int_0^\infty e^{tx} \frac{\lambda^\alpha}{\Gamma(\alpha)} x^{\alpha-1} e^{-\lambda x}\, dx \\ &= \frac{\lambda^\alpha}{\Gamma(\alpha)} \int_0^\infty x^{\alpha-1} e^{-(\lambda-t)x}\, dx \\ &= \frac{\lambda^\alpha}{\Gamma(\alpha)} \frac{\Gamma(\alpha)}{(\lambda - t)^\alpha} \end{aligned}$$

for $-\infty < t < \lambda$. The integral diverges for $\lambda \le t < \infty$. Thus

$$M_X(t) = \left(\frac{\lambda}{\lambda - t}\right)^\alpha, \qquad -\infty < t < \lambda. \tag{2}$$

Suppose now that X is a discrete random variable, all of whose possible values are nonnegative integers. Then

$$M_X(t) = \sum_{n=0}^{\infty} e^{nt} P(X = n).$$

In Chapter 3 we defined the probability generating function for such random variables as

$$\Phi_X(t) = \sum_{n=0}^{\infty} t^n P(X = n).$$

From these two formulas it is clear that

$$M_X(t) = \Phi_X(e^t). \tag{3}$$

Formula (3) allows us to determine the moment generating function directly from the probability generating function. For example, if X has a binomial distribution with parameters n and p, then as was shown in Example 16 of Chapter 3,

$$\Phi_X(t) = (pt + 1 - p)^n.$$

It follows immediately that

$$M_X(t) = (pe^t + 1 - p)^n.$$

Similarly, if X has a Poisson distribution with parameter λ, then according to Example 18 of Chapter 3,

$$\Phi_X(t) = e^{\lambda(t-1)}.$$

Consequently,

$$M_X(t) = e^{\lambda(e^t - 1)}.$$

Of course, in these two examples $M_X(t)$ could also easily be obtained directly from the definition of the moment generating function.

If X and Y are independent random variables, then e^{tX} and e^{tY} are also independent. Consequently

$$\begin{aligned} M_{X+Y}(t) = Ee^{t(X+Y)} &= Ee^{tX}e^{tY} = Ee^{tX}Ee^{tY} \\ &= M_X(t)M_Y(t). \end{aligned}$$

It follows easily that if $X_1, \ldots, X_n$ are independent and identically distributed, then

$$M_{X_1 + \cdots + X_n}(t) = (M_{X_1}(t))^n. \tag{4}$$

In order to see why $M_X(t)$ is called the moment generating function we write

$$M_X(t) = Ee^{tX} = E \sum_{n=0}^{\infty} \frac{t^n X^n}{n!}.$$

Suppose $M_X(t)$ is finite on $-t_0 < t < t_0$ for some positive number t_0. In this case one can show that in the last expression for $M_X(t)$ it is permissible to interchange the order of expectation and summation. In other words

$$M_X(t) = \sum_{n=0}^{\infty} \frac{EX^n}{n!} t^n \tag{5}$$

for $-t_0 < t < t_0$. In particular, if $M_X(t)$ is finite for all t, then (5) holds for all t. The Taylor series for $M_X(t)$ is

$$M_X(t) = \sum_{n=0}^{\infty} \frac{t^n}{n!} \frac{d^n}{dt^n} M_X(t) \bigg|_{t=0}. \tag{6}$$

By comparing the coefficients of t^n in (5) and (6), we see that

$$EX^n = \frac{d^n}{dt^n} M_X(t) \bigg|_{t=0}. \tag{7}$$

Example 3. Let X be normally distributed with mean 0 and variance σ^2. Use moment generating functions to find the moments of X.

Observe first from (1) that

$$\begin{aligned} M_X(t) = e^{\sigma^2 t^2/2} &= \sum_{n=0}^{\infty} \left(\frac{\sigma^2 t^2}{2}\right)^n \frac{1}{n!} \\ &= \sum_{n=0}^{\infty} \frac{\sigma^{2n}}{2^n n!} t^{2n}. \end{aligned}$$

Thus the odd moments of X are all zero, and the even moments are given by

$$\frac{EX^{2n}}{(2n)!} = \frac{\sigma^{2n}}{2^n n!}$$

or

$$EX^{2n} = \frac{\sigma^{2n}(2n)!}{2^n n!}.$$

This agrees with the result obtained in Chapter 7.

This example can also be used to illustrate (7). Since

$$\frac{d}{dt} e^{\sigma^2 t^2/2} = \sigma^2 t e^{\sigma^2 t^2/2}$$

and

$$\frac{d^2}{dt^2} e^{\sigma^2 t^2/2} = (\sigma^2 + \sigma^4 t^2) e^{\sigma^2 t^2/2},$$

it follows that

$$\left. \frac{d}{dt} e^{\sigma^2 t^2/2} \right|_{t=0} = 0$$

and

$$\left. \frac{d^2}{dt^2} e^{\sigma^2 t^2/2} \right|_{t=0} = \sigma^2,$$

which are just the first two moments of X.

8.2. Characteristic functions

The *characteristic function* of a random variable X is defined as

$$\varphi_X(t) = Ee^{itX}, \qquad -\infty < t < \infty,$$

where $i = \sqrt{-1}$. Characteristic functions are slightly more complicated than moment generating functions in that they involve complex numbers. They have, however, two important advantages over moment generating functions. First, $\varphi_X(t)$ is finite for all random variables X and all real numbers t. Secondly, the distribution function of X and usually the density function, if it exists, can be obtained from the characteristic function by means of an "inversion formula." Using properties of characteristic functions we will be able to prove both the Weak Law of Large Numbers and the Central Limit Theorem, which we would not be able to do with moment generating functions.

Before discussing characteristic functions we will first briefly summarize some required facts involving complex variables.

We can write any complex number z in the form $z = x + iy$, where x and y are real numbers. The absolute value $|z|$ of such a complex number is defined by $|z| = (x^2 + y^2)^{1/2}$. The distance between two complex numbers z_1 and z_2 is defined to be $|z_1 - z_2|$.

If a function of a real variable has a power series expansion with a positive radius of convergence, we can use that power series to define a corresponding function of a complex variable. Thus we define

$$e^z = \sum_{n=0}^{\infty} \frac{z^n}{n!}$$

for any complex number z. The relation

$$e^{z_1+z_2} = e^{z_1}e^{z_2}$$

remains valid for all complex numbers z_1 and z_2. Letting $z = it$, where t is a real number, we see that

$$\begin{aligned} e^{it} &= \sum_{n=0}^{\infty} \frac{(it)^n}{n!} \\ &= \left(1 + it - \frac{t^2}{2} - \frac{it^3}{3!} + \frac{t^4}{4!} + \frac{it^5}{5!} - \cdots\right) \\ &= \left(1 - \frac{t^2}{2!} + \frac{t^4}{4!} - \cdots\right) + i\left(t - \frac{t^3}{3!} + \frac{t^5}{5!} - \cdots\right). \end{aligned}$$

Since the two power series in the last expression are those of $\cos t$ and $\sin t$, it follows that

$$e^{it} = \cos t + i \sin t. \tag{8}$$

Using the fact that $\cos(-t) = \cos t$ and $\sin(-t) = -\sin t$, we see that

$$e^{-it} = \cos t - i \sin t.$$

From these formulas we can solve for $\cos t$ and $\sin t$, obtaining

$$\cos t = \frac{e^{it} + e^{-it}}{2} \quad \text{and} \quad \sin t = \frac{e^{it} - e^{-it}}{2i}. \tag{9}$$

It also follows from (8) that

$$|e^{it}| = (\cos^2 t + \sin^2 t)^{1/2} = 1.$$

If $f(t)$ and $g(t)$ are real-valued functions of t, then $h(t) = f(t) + ig(t)$ defines a complex-valued function of t. We can differentiate $h(t)$ by differentiating $f(t)$ and $g(t)$ separately; that is,

$$h'(t) = f'(t) + ig'(t),$$

provided that $f'(t)$ and $g'(t)$ exist. Similarly we define

$$\int_a^b h(t)\, dt = \int_a^b f(t)\, dt + i \int_a^b g(t)\, dt,$$

provided that the indicated integrals involving f and g exist. The formula

$$\frac{d}{dt} e^{ct} = ce^{ct}$$

is valid for any complex constant c. The fundamental theorem of calculus continues to hold and, in particular, if c is a nonzero complex constant, then

$$\int_a^b e^{ct}\,dt = \frac{e^{cb} - e^{ca}}{c}.$$

A complex-valued random variable Z can be written in the form $Z = X + iY$, where X and Y are real-valued random variables. Its expectation EZ is defined as

$$EZ = E(X + iY) = EX + iEY$$

whenever EX and EY are well defined. Just as for real-valued random variables, Z has finite expectation if and only if $E|Z| < \infty$, and in that case

$$|EZ| \leq E|Z|.$$

The formula

$$E(a_1Z_1 + a_2Z_2) = a_1EZ_1 + a_2EZ_2$$

is valid whenever a_1 and a_2 are complex constants and Z_1 and Z_2 are complex-valued random variables having finite expectation.

We will let X and Y, with or without subscripts, continue to denote real-valued random variables. Thus in the phrase "let X be a random variable . . ." it is understood that X is real-valued.

Suppose now that X is a random variable and t is a constant (we reserve the symbol t for real constants). Then $|e^{itX}| = 1$, so that e^{itX} has finite expectation and the characteristic function $\varphi_X(t)$, $-\infty < t < \infty$, given by

$$\varphi_X(t) = Ee^{itX}, \qquad -\infty < t < \infty,$$

is well defined. We see that $\varphi_X(0) = Ee^0 = E1 = 1$ and, for $-\infty < t < \infty$,

$$|\varphi_X(t)| = |Ee^{itX}| \leq E|e^{itX}| = E1 = 1.$$

The reason characteristic functions are finite for all t whereas moment generating functions are not finite in general is that e^{it}, $-\infty < t < \infty$, is bounded while e^t, $-\infty < t < \infty$, is unbounded.

Example 4. Let X be a random variable taking on the value a with probability one. Then

$$\varphi_X(t) = Ee^{itX} = e^{ita}, \qquad -\infty < t < \infty.$$

In particular, if X takes on the value zero with probability one, then its characteristic function is identically equal to 1.

If X is a random variable and a and b are real constants, then

$$\begin{aligned}\varphi_{a+bX}(t) &= Ee^{it(a+bX)}\\ &= Ee^{ita}e^{ibtX}\\ &= e^{ita}Ee^{ibtX},\end{aligned}$$

and hence

(10) $$\varphi_{a+bX}(t) = e^{ita}\varphi_X(bt), \qquad -\infty < t < \infty.$$

Example 5. Let U be uniformly distributed on $(-1, 1)$. Then for $t \neq 0$

$$\varphi_U(t) = \int_{-1}^{1} e^{itu}\,\frac{1}{2}\,du$$

$$= \frac{1}{2}\,\frac{e^{itu}}{it}\bigg|_{-1}^{1}$$

$$= \frac{1}{2}\left(\frac{e^{it} - e^{-it}}{it}\right)$$

$$= \frac{\sin t}{t}.$$

For $a < b$ let

$$X = \frac{a+b}{2} + \left(\frac{b-a}{2}\right)U.$$

Then X is uniformly distributed on (a, b), and by (10) for $t \neq 0$

$$\varphi_X(t) = e^{it(a+b)/2}\,\frac{\sin((b-a)t/2)}{(b-a)t/2}.$$

Alternatively

$$\varphi_X(t) = \int_a^b e^{itx}\,\frac{1}{b-a}\,dx$$

$$= \frac{1}{b-a}\,\frac{e^{itx}}{it}\bigg|_a^b$$

$$= \frac{e^{ibt} - e^{iat}}{it(b-a)}.$$

It is easy to check by means of (9) that these two answers agree.

Example 6. Let X have an exponential distribution with parameter λ. Then

$$\varphi_X(t) = \int_0^\infty e^{itx}\lambda e^{-\lambda x}\,dx$$

$$= \lambda \int_0^\infty e^{-(\lambda - it)x}\,dx$$

$$= \frac{\lambda}{\lambda - it}\,e^{-(\lambda - it)x}\bigg|_\infty^0.$$

Since $\lim_{x\to\infty} e^{-\lambda x} = 0$ and e^{itx} is bounded in x, it follows that

$$\lim_{x\to\infty} e^{-(\lambda - it)x} = \lim_{x\to\infty} e^{-\lambda x}e^{itx} = 0.$$

Thus

$$\varphi_X(t) = \frac{\lambda}{\lambda - it}.$$

Suppose X and Y are independent random variables. Then e^{itX} and e^{itY} are also independent random variables; consequently

$$\varphi_{X+Y}(t) = Ee^{it(X+Y)} = Ee^{itX}e^{itY} = Ee^{itX}Ee^{itY}$$

and hence

(11) $$\varphi_{X+Y}(t) = \varphi_X(t)\varphi_Y(t), \qquad -\infty < t < \infty.$$

Formula (11) extends immediately to yield the fact that the characteristic function of the sum of a finite number of independent random variables is the product of the individual characteristic functions.

It can be shown that $\varphi_X(t)$ is a continuous function of t. Moreover, if X has finite nth moment, then $\varphi_X^{(n)}(t)$ exists, is continuous in t, and can be calculated as

$$\varphi_X^{(n)}(t) = \frac{d^n}{dt^n} Ee^{itX} = E\frac{d^n}{dt^n} e^{itX} = E(iX)^n e^{itX}.$$

In particular

(12) $$\varphi_X^{(n)}(0) = i^n EX^n.$$

We can attempt to expand $\varphi_X(t)$ into a power series according to the formula

(13) $$\varphi_X(t) = Ee^{itX} = E\sum_{n=0}^{\infty} \frac{(itX)^n}{n!} = \sum_{n=0}^{\infty} \frac{i^n EX^n}{n!} t^n.$$

Suppose that

$$M_X(t) = \sum_{n=0}^{\infty} \frac{EX^n}{n!} t^n$$

is finite on $-t_0 < t < t_0$ for some positive number t_0. Then (13) also holds on $-t_0 < t < t_0$.

Example 7. Let X be normally distributed with mean 0 and variance σ^2. Find $\varphi_X(t)$.

From Chapter 7 we know that $EX^n = 0$ for any odd positive integer n. Furthermore, if $n = 2k$ is an even integer, then

$$EX^n = EX^{2k} = \frac{\sigma^{2k}(2k)!}{2^k k!}.$$

Therefore

$$\varphi_X(t) = \sum_{k=0}^{\infty} \frac{i^{2k}EX^{2k}}{(2k)!} t^{2k} = \sum_{k=0}^{\infty} \frac{(-\sigma^2 t^2/2)^k}{k!} = e^{-\sigma^2 t^2/2}.$$

More generally let X be normally distributed with mean μ and variance σ^2. Then $Y = X - \mu$ is normally distributed with mean 0 and variance σ^2. Since $X = \mu + Y$ we see from Formula (10) and Example 7 that

$$\varphi_X(t) = e^{it\mu}e^{-\sigma^2t^2/2}, \qquad -\infty < t < \infty. \tag{14}$$

Let X be a random variable whose moment generating function $M_X(t)$ is finite on $-t_0 < t < t_0$ for some positive number t_0. Since

$$M_X(t) = Ee^{tX}$$

and

$$\varphi_X(t) = Ee^{itX},$$

we would expect that

$$\varphi_X(t) = M_X(it). \tag{15}$$

In other words, we would expect that if we replace t by it in the formula for the moment generating function, we will obtain the corresponding formula for the characteristic function. This is indeed the case, but a thorough understanding of the issues involved requires a sophisticated concept (analytic continuation) from complex variable theory.

As an example of (15), let X be normally distributed with mean μ and variance σ^2. Then as we have already seen

$$M_X(t) = e^{\mu t}e^{\sigma^2t^2/2}$$

and hence

$$\begin{aligned} M_X(it) &= e^{\mu(it)}e^{\sigma^2(it)^2/2} \\ &= e^{i\mu t}e^{-\sigma^2t^2/2} \end{aligned}$$

which by (14) is $\varphi_X(t)$.

8.3. Inversion formulas and the Continuity Theorem

Let X be an integer-valued random variable. Its characteristic function is given by

$$\varphi_X(t) = \sum_{-\infty}^{\infty} e^{ijt}f_X(j).$$

One of the most useful properties of $\varphi_X(t)$ is that it can be used to calculate $f_X(k)$. Specifically we have the "inversion formula"

$$f_X(k) = \frac{1}{2\pi}\int_{-\pi}^{\pi} e^{-ikt}\varphi_X(t)\,dt. \tag{16}$$

In order to verify (16) we write the right side of this formula as

$$\frac{1}{2\pi}\int_{-\pi}^{\pi} e^{-ikt}\left[\sum_{-\infty}^{\infty} e^{ijt}f_X(j)\right] dt.$$

A theorem in integration theory justifies interchanging the order of integration and summation to yield the expression

$$\sum_{-\infty}^{\infty} f_X(j) \frac{1}{2\pi} \int_{-\pi}^{\pi} e^{i(j-k)t} \, dt.$$

In order to complete the proof of (16) we must show that the last expression equals $f_X(k)$. To do so it is enough to show that

$$\frac{1}{2\pi} \int_{-\pi}^{\pi} e^{i(j-k)t} \, dt = \begin{cases} 1 & \text{if } j = k, \\ 0 & \text{if } j \neq k. \end{cases} \tag{17}$$

Formula (17) is obvious when $j = k$, for in that case $e^{i(j-k)t} = 1$ for all t. If $j \neq k$, then

$$\begin{aligned} \frac{1}{2\pi} \int_{-\pi}^{\pi} e^{i(j-k)t} \, dt &= \frac{e^{i(j-k)t}\big|_{-\pi}^{\pi}}{2\pi i(j-k)} \\ &= \frac{e^{i(j-k)\pi} - e^{-i(j-k)\pi}}{2\pi i(j-k)} \\ &= \frac{\sin (j-k)\pi}{\pi(j-k)} = 0, \end{aligned}$$

since $\sin m\pi = 0$ for all integers m. This completes the proof of (17) and hence also of (16).

Example 8. Let $X_1, X_2, \ldots, X_n$ be independent, identically distributed integer-valued random variables and set $S_n = X_1 + \cdots + X_n$. Then $\varphi_{S_n}(t) = (\varphi_{X_1}(t))^n$, and consequently by (16)

$$f_{S_n}(k) = \frac{1}{2\pi} \int_{-\pi}^{\pi} e^{-ikt} (\varphi_{X_1}(t))^n \, dt. \tag{18}$$

Formula (18) is the basis of almost all methods of analyzing the behavior of $f_{S_n}(k)$ for large values of n and, in particular, the basis for the proof of the "local" Central Limit Theorem discussed in Chapter 7.

There is also an analog of (16) for continuous random variables. Let X be a random variable whose characteristic function $\varphi_X(t)$ is integrable, that is,

$$\int_{-\infty}^{\infty} |\varphi_X(t)| \, dt < \infty.$$

It can be shown that in this case X is a continuous random variable having a density f_X given by

$$f_X(x) = \frac{1}{2\pi} \int_{-\infty}^{\infty} e^{-ixt} \varphi_X(t) \, dt. \tag{19}$$

Example 9. Let X be normally distributed with mean 0 and variance σ^2. We will show directly that (19) is valid for such a random variable.

From Example 7 we know that X has characteristic function $\varphi_X(t) = e^{-\sigma^2 t^2/2}$. Thus by the definition of characteristic functions,

$$e^{-\sigma^2 t^2/2} = \int_{-\infty}^{\infty} e^{itx} \frac{1}{\sigma\sqrt{2\pi}} e^{-x^2/2\sigma^2}\, dx.$$

If we replace t by $-t$ and σ by $1/\sigma$ in this formula it becomes

$$e^{-t^2/2\sigma^2} = \int_{-\infty}^{\infty} e^{-itx} \frac{\sigma}{\sqrt{2\pi}} e^{-\sigma^2 x^2/2}\, dx$$

or equivalently,

$$\frac{1}{\sigma\sqrt{2\pi}} e^{-t^2/2\sigma^2} = \frac{1}{2\pi} \int_{-\infty}^{\infty} e^{-itx} e^{-\sigma^2 x^2/2}\, dx.$$

Finally, if we interchange the role of the symbols x and t in the last equation we obtain

$$\frac{1}{\sigma\sqrt{2\pi}} e^{-x^2/2\sigma^2} = \frac{1}{2\pi} \int_{-\infty}^{\infty} e^{-itx} e^{-\sigma^2 t^2/2}\, dt,$$

which is just (19) in this special case.

Let X now denote any random variable. Let Y be a random variable that is independent of X and has the standard normal distribution, and let c denote a positive constant. Then $X + cY$ has the characteristic function

$$\varphi_X(t) e^{-c^2 t^2/2}.$$

Since $\varphi_X(t)$ is bounded in absolute value by 1 and $e^{-c^2t^2/2}$ is integrable, it follows that $X + cY$ has an integrable characteristic function. Consequently (19) is applicable and $X + cY$ is a continuous random variable having a density given by

$$f_{X+cY}(x) = \frac{1}{2\pi} \int_{-\infty}^{\infty} e^{-itx} \varphi_X(t) e^{-c^2t^2/2}\, dt.$$

If we integrate both sides of this equation over $a \le x \le b$ and interchange the order of integration, we conclude that

$$\begin{aligned} P(a \le X + cY \le b) &= \frac{1}{2\pi} \int_a^b \left(\int_{-\infty}^{\infty} e^{-itx} \varphi_X(t) e^{-c^2t^2/2}\, dt \right) dx \\ &= \frac{1}{2\pi} \int_{-\infty}^{\infty} \left(\int_a^b e^{-itx}\, dx \right) \varphi_X(t) e^{-c^2t^2/2}\, dt \end{aligned}$$

or

$$(20) \qquad P(a \le X + cY \le b) = \frac{1}{2\pi} \int_{-\infty}^{\infty} \left(\frac{e^{-ibt} - e^{-iat}}{-it} \right) \varphi_X(t) e^{-c^2t^2/2}\, dt.$$

The importance of (20) is that it holds for an arbitrary random variable X.

The right side of (20) depends on X only through $\varphi_X(t)$. By using this fact and letting $c \to 0$ in (20), one can show that the distribution function of X is determined by its characteristic function. This result is known as a "uniqueness theorem" and can be stated as follows:

Theorem 1 *If two random variables have the same characteristic function, they have the same distribution function.*

Example 10. Use the uniqueness theorem to show that the sum of two independent normally distributed random variables is itself normally distributed.

Let X and Y be independent and distributed respectively according to $n(\mu_1, \sigma_1^2)$ and $n(\mu_2, \sigma_2^2)$. Then

$$\varphi_X(t) = e^{i\mu_1 t}e^{-\sigma_1^2 t^2/2}$$

and

$$\varphi_Y(t) = e^{i\mu_2 t}e^{-\sigma_2^2 t^2/2}.$$

Consequently

$$\varphi_{X+Y}(t) = e^{i(\mu_1+\mu_2)t}e^{-(\sigma_1^2+\sigma_2^2)t^2/2}.$$

Thus the characteristic function of $X + Y$ is the same as that of a random variable having a normal distribution with mean $\mu_1 + \mu_2$ and variance $\sigma_1^2 + \sigma_2^2$. By the uniqueness theorem $X + Y$ must have that normal distribution.

The most important application of the inversion formula (20) is that it can be used to derive the next result, which is basic to the proof of the Weak Law of Large Numbers and the Central Limit Theorem.

Theorem 2 *Let X_n, $n \geq 1$, and X be random variables such that*

$$\lim_{n \to \infty} \varphi_{X_n}(t) = \varphi_X(t), \qquad -\infty < t < \infty. \tag{21}$$

Then

$$\lim_{n \to \infty} F_{X_n}(x) = F_X(x) \tag{22}$$

at all points x where F_X is continuous.

This theorem states that convergence of characteristic functions implies convergence of the corresponding distribution functions or, in other words, that distribution functions "depend continuously" on their characteristic functions. For this reason Theorem 2 is commonly known as the "Continuity Theorem."

The proof of this theorem is rather involved. We will not present the details of the proof, but we will indicate briefly some of the main ideas of one method of proof.

We first choose a random variable Y that has the standard normal distribution and is independent of each of the random variables X_n, $n \geq 1$.

Let $a < b$ and let c be a positive constant. Then by the inversion formula (20)

$$(23) \qquad P(a \le X_n + cY \le b) = \frac{1}{2\pi} \int_{-\infty}^{\infty} \left(\frac{e^{-ibt} - e^{-iat}}{-it} \right) \varphi_{X_n}(t) e^{-c^2t^2/2} \, dt$$

and

$$(24) \qquad P(a \le X + cY \le b) = \frac{1}{2\pi} \int_{-\infty}^{\infty} \left(\frac{e^{-ibt} - e^{-iat}}{-it} \right) \varphi_X(t) e^{-c^2t^2/2} \, dt.$$

By assumption $\varphi_{X_n}(t) \to \varphi_X(t)$ as $n \to \infty$. It follows from this by a theorem in integration theory that the right side of (23) converges to the right side of (24) as $n \to \infty$. Consequently

$$(25) \qquad \lim_{n \to \infty} P(a \le X_n + cY \le b) = P(a \le X + cY \le b).$$

There are two more steps to the proof of the theorem. First one must show (by letting $a \to -\infty$ in (25)) that

$$(26) \qquad \lim_{n \to \infty} P(X_n + cY \le b) = P(X + cY \le b).$$

Finally, one must show (by letting $c \to 0$ in (26)) that

$$\lim_{n \to \infty} P(X_n \le b) = P(X \le b)$$

whenever $P(X = b) = 0$. The last result is equivalent to the conclusion of the theorem.

8.4. The Weak Law of Large Numbers and the Central Limit Theorem

In this section we will use the Continuity Theorem to prove the two important theorems in probability theory stated in the title to this section. Both theorems were discussed without proof in earlier chapters. In order to prove these theorems we first need to study the asymptotic behavior of $\log \varphi_X(t)$ near $t = 0$.

Let z be a complex number such that $|z - 1| < 1$. We can define $\log z$ by means of the power series

$$\log z = (z - 1) - \frac{(z-1)^2}{2} + \frac{(z-1)^3}{3} - \cdots$$

(for $|z - 1| \ge 1$, other definitions of $\log z$ are needed). With this definition we have the usual properties that $\log 1 = 0$,

$$e^{\log z} = z, \qquad |z - 1| < 1,$$

and if $h(t)$, $a < t < b$, is a differentiable complex-valued function such that $|h(t) - 1| < 1$, then

$$\frac{d}{dt} \log h(t) = \frac{h'(t)}{h(t)}.$$

Let X be a random variable having characteristic function $\varphi_X(t)$. Then $\varphi_X(t)$ is continuous and $\varphi_X(0) = 1$. Thus $\log \varphi_X(t)$ is well defined for t near 0 and $\log \varphi_X(0) = 0$.

Suppose now that X has finite mean μ. Then $\varphi_X(t)$ is differentiable and, by (12), $\varphi'_X(0) = i\mu$. Hence

$$\begin{aligned}\lim_{t\to 0} \frac{\log \varphi_X(t)}{t} &= \lim_{t\to 0} \frac{\log \varphi_X(t) - \log \varphi_X(0)}{t - 0} \\ &= \frac{d}{dt} \log \varphi_X(t)\Big|_{t=0} \\ &= \frac{\varphi'_X(0)}{\varphi_X(0)} \\ &= i\mu.\end{aligned}$$

Consequently,

$$\lim_{t\to 0} \frac{\log \varphi_X(t) - i\mu t}{t} = 0. \tag{27}$$

Suppose now that X also has finite variance σ^2. Then $\varphi_X(t)$ is twice differentiable and by (12)

$$\varphi''_X(0) = -EX^2 = -(\mu^2 + \sigma^2).$$

We can apply l'Hôspital's rule to obtain

$$\begin{aligned}\lim_{t\to 0} \frac{\log \varphi_X(t) - i\mu t}{t^2} &= \lim_{t\to 0} \frac{\dfrac{\varphi'_X(t)}{\varphi_X(t)} - i\mu}{2t} \\ &= \lim_{t\to 0} \frac{\varphi'_X(t) - i\mu\varphi_X(t)}{2t\varphi_X(t)} \\ &= \lim_{t\to 0} \frac{\varphi'_X(t) - i\mu\varphi_X(t)}{2t}.\end{aligned}$$

By a second application of l'Hôspital's rule we see that

$$\begin{aligned}\lim_{t\to 0} \frac{\log \varphi_X(t) - i\mu t}{t^2} &= \frac{\varphi''_X(0) - i\mu\varphi'_X(0)}{2} \\ &= \frac{-(\mu^2 + \sigma^2) - (i\mu)^2}{2} \\ &= \frac{-\mu^2 - \sigma^2 + \mu^2}{2}.\end{aligned}$$

In other words

$$\lim_{t\to 0} \frac{\log \varphi_X(t) - i\mu t}{t^2} = -\frac{\sigma^2}{2}. \tag{28}$$

Theorem 3 (Weak Law of Large Numbers). *Let $X_1, X_2, \ldots$ be independent, identically distributed random variables having finite mean μ and set $S_n = X_1 + \cdots + X_n$. Then for any $\varepsilon > 0$*

$$\lim_{n \to \infty} P\left(\left|\frac{S_n}{n} - \mu\right| > \varepsilon\right) = 0. \tag{29}$$

Proof. The characteristic function of

$$\frac{S_n}{n} - \mu = \frac{X_1 + \cdots + X_n}{n} - \mu$$

is

$$e^{-i\mu t}(\varphi_{X_1}(t/n))^n.$$

Let t be fixed. Then for n sufficiently large, t/n is close enough to zero so that $\log \varphi_{X_1}(t/n)$ is well defined and

$$e^{-i\mu t}(\varphi_{X_1}(t/n))^n = \exp\left[n(\log \varphi_{X_1}(t/n) - i\mu(t/n))\right]. \tag{30}$$

We claim next that

$$\lim_{n \to \infty} n(\log \varphi_{X_1}(t/n) - i\mu(t/n)) = 0. \tag{31}$$

Equation (31) is obvious for $t = 0$ since $\log \varphi_{X_1}(0) = \log 1 = 0$. If $t \neq 0$ we can write the left side of (31) as

$$t \lim_{n \to \infty} \frac{\log \varphi_X(t/n) - i\mu(t/n)}{t/n}.$$

But $t/n \to 0$ as $n \to \infty$, so the last limit is 0 by (27). This completes the proof of (31). It follows from (30) and (31) that the characteristic function of

$$\frac{S_n}{n} - \mu$$

approaches 1 as $n \to \infty$. Now 1 is the characteristic function of a random variable X such that $P(X = 0) = 1$. The distribution function of X is given by

$$F_X(x) = \begin{cases} 1 & \text{if } x \geq 0, \\ 0 & \text{if } x < 0. \end{cases}$$

This distribution function is continuous everywhere except at $x = 0$. Choose $\varepsilon > 0$. Then by the Continuity Theorem,

$$\lim_{n \to \infty} P\left(\frac{S_n}{n} - \mu \leq -\varepsilon\right) = F_X(-\varepsilon) = 0 \tag{32}$$

and

$$\lim_{n \to \infty} P\left(\frac{S_n}{n} - \mu \leq \varepsilon\right) = F_X(\varepsilon) = 1.$$

The last result implies that

$$\lim_{n\to\infty} P\left(\frac{S_n}{n} - \mu > \varepsilon\right) = 0,$$

which together with (32) implies that (29) holds as desired. ∎

For the next theorem it is necessary to recall that $\Phi(x)$ denotes the standard normal distribution function given by

$$\Phi(x) = \int_{-\infty}^{x} \frac{1}{\sqrt{2\pi}} e^{-y^2/2}\, dy, \qquad -\infty < x < \infty.$$

We recall also that this distribution function is continuous at all values of x.

Theorem 4 (Central Limit Theorem). *Let $X_1, X_2, \ldots$ be independent, identically distributed random variables each having mean μ and finite nonzero variance σ^2. Then*

$$\lim_{n\to\infty} P\left(\frac{S_n - n\mu}{\sigma\sqrt{n}} \le x\right) = \Phi(x), \qquad -\infty < x < \infty.$$

Proof. Set

$$S_n^* = \frac{S_n - n\mu}{\sigma\sqrt{n}}.$$

Then for t fixed and n sufficiently large,

$$\begin{aligned} \varphi_{S_n^*}(t) &= e^{-in\mu t/\sigma\sqrt{n}} \varphi_{S_n}(t/\sigma\sqrt{n}) \\ &= e^{-in\mu t/\sigma\sqrt{n}} (\varphi_{X_1}(t/\sigma\sqrt{n}))^n, \end{aligned}$$

or

$$\varphi_{S_n^*}(t) = \exp\left[n(\log \varphi_{X_1}(t/\sigma\sqrt{n}) - i\mu(t/\sigma\sqrt{n}))\right]. \tag{33}$$

We claim next that

$$\lim_{n\to\infty} n(\log \varphi_{X_1}(t/\sigma\sqrt{n}) - i\mu(t/\sigma\sqrt{n})) = -\frac{t^2}{2}. \tag{34}$$

If $t = 0$, then both sides of (34) equal zero and (34) clearly holds. If $t \ne 0$ we can write the left side of (34) as

$$\frac{t^2}{\sigma^2} \lim_{n\to\infty} \frac{\log \varphi_{X_1}(t/\sigma\sqrt{n}) - i\mu(t/\sigma\sqrt{n})}{(t/\sigma\sqrt{n})^2},$$

which by (28) equals

$$\frac{t^2}{\sigma^2}\left(-\frac{\sigma^2}{2}\right) = -\frac{t^2}{2}.$$

Thus (34) holds for all t. It follows from (33) and (34) that

$$\lim_{n\to\infty} \varphi_{S_n^*}(t) = e^{-t^2/2}, \qquad -\infty < t < \infty.$$

According to Example 7, $e^{-t^2/2}$ is the characteristic function of a random variable X having the standard normal distribution function $\Phi(x)$. Thus by the Continuity Theorem

$$\lim_{n\to\infty} P(S_n^* \le x) = \Phi(x), \qquad -\infty < x < \infty,$$

which is the desired conclusion. ∎

Exercises

1 Let X be uniformly distributed on (a, b). Find $M_X(t)$.

2 Express the moment generating function of $Y = a + bX$ in terms of $M_X(t)$ (here a and b are constants).

3 Let X have a Poisson distribution with parameter λ. Use moment generating functions to find the mean and variance of X.

4 Let X have a negative binomial distribution with parameters α and p.
(a) Find the moment generating function of X.
(b) Use this moment generating function to find the mean and variance of X.

5 Let X be a continuous random variable having the density $f_X(x) = (1/2)e^{-|x|}$, $-\infty < x < \infty$.
(a) Show that $M_X(t) = 1/(1 - t^2)$, $-1 < t < 1$.
(b) Use this moment generating function to find a formula for EX^{2n} (note that the odd moments of X are all zero).

6 Let X have a binomial distribution with parameters n and p.
(a) Find $dM_X(t)/dt$ and $d^2M_X(t)/dt^2$.
(b) Use (a) and Formula (7) to compute the mean and variance of X.

7 Let $X_1, \ldots, X_n$ be independent, identically distributed random variables such that $M_{X_1}(t)$ is finite for all t. Use moment generating functions to show that

$$E(X_1 + \cdots + X_n)^3 = nEX_1^3 + 3n(n-1)EX_1^2EX_1 + n(n-1)(n-2)(EX_1)^3.$$

Hint: Find $(d^3/dt^3)(M_{X_1}(t))^n|_{t=0}$.

8 Let X be a random variable such that $M_X(t)$ is finite for all t. Use the same argument as in the proof of Chebyshev's Inequality to conclude that

$$P(X \ge x) \le e^{-tx}M_X(t), \qquad t \ge 0.$$

It follows that

$$P(X \geq x) \leq \min_{t \geq 0} e^{-tx} M_X(t),$$

provided that $e^{-tx}M_X(t)$ has a minimum on $0 \leq t < \infty$.

9 Let X have a gamma distribution with parameters α and λ. Use the result of Exercise 8 to show that $P(X \geq 2\alpha/\lambda) \leq (2/e)^\alpha$.

10 Let X have a Poisson distribution with parameter λ. Find $\varphi_X(t)$.

11 Let X have a geometric distribution with parameter p. Find $\varphi_X(t)$.

12 Let $X_1, X_2, \ldots, X_n$ be independent random variables each having a geometric distribution with parameter p. Find the characteristic function of $X = X_1 + \cdots + X_n$.

13 Let $X_1, X_2, \ldots, X_n$ be independent random variables each having an exponential distribution with parameter λ. Find the characteristic function of $X = X_1 + \cdots + X_n$.

14 Let X be a discrete random variable all of whose possible values are nonnegative integers. What relationship should we expect to hold between the characteristic function of X and the probability generating function of X (recall Formulas (3) and (15))?

15 Let X be any random variable.
(a) Show that $\varphi_X(t) = E \cos tX + iE \sin tX$.
(b) Show that $\varphi_{-X}(t) = E \cos tX - iE \sin tX$.
(c) Show that $\varphi_{-X}(t) = \varphi_X(-t)$.

16 Let X be a symmetric random variable, that is, such that X and $-X$ have the same distribution function.
(a) Show that $E \sin tX = 0$ and that $\varphi_X(t)$ is real-valued.
(b) Show that $\varphi_X(-t) = \varphi_X(t)$.

17 Let X and Y be independent, identically distributed random variables. Show that $\varphi_{X-Y}(t) = |\varphi_X(t)|^2$. *Hint:* Use Exercise 15.

18 Let X be a random variable such that $\varphi_X(t)$ is real-valued.
(a) Show that X and $-X$ have the same characteristic function (use Exercise 15).
(b) Why does it follow that X and $-X$ have the same distribution function?

19 Let X be a continuous random variable having the density $f_X(x) = (1/2)e^{-|x|}$, $-\infty < x < \infty$.
(a) Show that $\varphi_X(t) = 1/(1 + t^2)$.
(b) Use (a) and the inversion formula (19) to conclude that

$$e^{-|x|} = \int_{-\infty}^{\infty} e^{-ixt} \frac{1}{\pi(1 + t^2)} \, dt.$$

(c) Show by using (b) that

$$e^{-|x|} = \int_{-\infty}^{\infty} e^{ixt} \frac{1}{\pi(1 + t^2)} \, dt.$$

20 Let X be a random variable having the Cauchy density

$$f_X(x) = \frac{1}{\pi(1 + x^2)}, \qquad -\infty < x < \infty.$$

Show that $\varphi_X(t) = e^{-|t|}$, $-\infty < t < \infty$. *Hint:* Interchange the role of x and t in Exercise 19.

21 Let X and Y be independent random variables having the Cauchy density.
(a) Find the characteristic functions of $X + Y$ and of $(X + Y)/2$.
(b) Why does it follow that $(X + Y)/2$ also has the Cauchy density?

22 Extend the result of Exercise 21 by showing that if $X_1, X_2, \ldots, X_n$ are independent random variables each having the Cauchy density, then $(X_1 + \cdots + X_n)/n$ also has the Cauchy density.

23 For $\lambda > 0$ let X_λ be a random variable having a Poisson distribution with parameter λ.
(a) Use arguments similar to those used in proving the Central Limit Theorem to show that for $-\infty < t < \infty$,

$$\lim_{\lambda \to \infty} Ee^{it(X_\lambda - \lambda)/\sqrt{\lambda}} = \lim_{\lambda \to \infty} \exp\left[\lambda(e^{it/\sqrt{\lambda}} - 1 - it/\sqrt{\lambda})\right] = e^{-t^2/2}.$$

(b) What conclusion should follow from (a) by an appropriate modification of the Continuity Theorem?

9 Random Walks and Poisson Processes

In this chapter we discuss two elementary but important examples of stochastic processes. A *stochastic process* may be defined as any collection of random variables. Usually, however, in referring to a stochastic process we have in mind a process that has enough additional structure so that interesting and useful results can be obtained. This is certainly true of the two examples treated in this chapter. The material on Poisson processes, our second example, does not depend on the first two sections where we discuss random walks.

9.1. Random walks

Consider a sequence of games such that during the nth game a random variable X_n is observed and any player playing the nth game receives the amount X_n from the "house" (of course, if $X_n < 0$ the player actually pays $-X_n$ to the house).

Let us follow the progress of a player starting out with initial capital x. Let S_n, $n \geq 0$, denote his capital after n games. Then $S_0 = x$ and

$$S_n = x + X_1 + \cdots + X_n, \qquad n \geq 1.$$

The collection of random variables $S_0, S_1, S_2, \ldots$ is an example of a stochastic process. In order to get interesting results we will assume that the random variables $X_1, X_2, \ldots$ are independent and identically distributed. Under this assumption the process $S_0, S_1, \ldots$ is called a *random walk*.

We will further assume that the X_k's have finite mean μ. If a player plays the first n games, his expected capital at the conclusion of the nth game is

$$ES_n = x + n\mu. \tag{1}$$

Suppose, however, that the player chooses numbers $a \leq x$ and $b \geq x$ and makes a bargain with himself to quit playing when his capital becomes not greater than a or not less than b. Then the number of times T that he will play the game is a random variable defined by

$$T = \min (n \geq 0 \mid S_n \leq a \quad \text{or} \quad S_n \geq b).$$

In order to guarantee that $S_n \leq a$ or $S_n \geq b$ for some n, we assume that

$$P(X_k = 0) < 1. \tag{2}$$

It is possible to prove that the random variable T is finite (with probability 1) and, in fact, $P(T > n)$ decreases exponentially as $n \to \infty$. This means that for some positive constants M and $c < 1$,

$$P(T > n) < Mc^n, \qquad n = 0, 1, 2, \ldots. \tag{3}$$

The proof of (3) is not difficult but will be omitted to allow room for results of much greater interest. From (3) and Theorem 5 of Chapter 4, it follows that ET and all higher moments of T are finite.

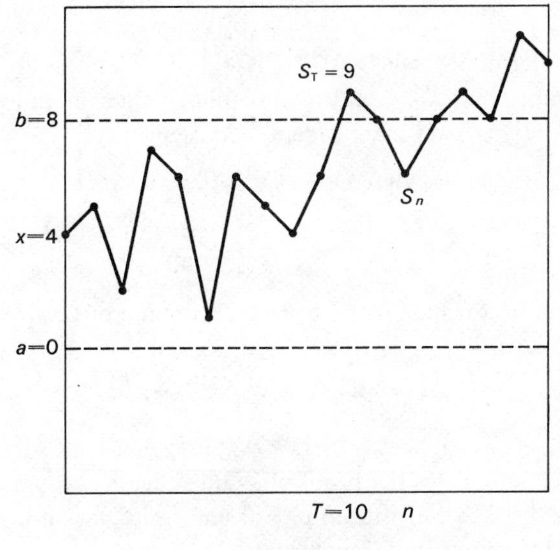

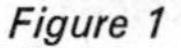

Figure 1

If the player quits playing after the Tth game, his capital will be S_T (see Figure 1). A famous identity due to Abraham Wald relates the expected capital when the player quits playing to the expected number of times he plays the game. Specifically, Wald's identity asserts that

$$ES_T = x + \mu ET. \tag{4}$$

Wald's identity is remarkably similar to (1).

In proving Wald's identity it is convenient to introduce a new notation. Let A be any event. By 1_A we mean the indicator random variable of A, that is, the random variable that is 1 if A occurs and 0 if A does not occur. By definition $1_A + 1_{A^c} = 1$. Using this notation we can write

$$S_T = x + \sum_{j=1}^{T} X_j = x + \sum_{j=1}^{\infty} X_j 1_{\{T \geq j\}}.$$

Since the complement of the event $\{T \geq j\}$ is the event $\{T < j\}$, we see that

$$S_T = x + \sum_{j=1}^{\infty} X_j(1 - 1_{\{T<j\}}), \tag{5}$$

and hence

$$ES_T = x + E \sum_{j=1}^{\infty} X_j(1 - 1_{\{T<j\}}). \tag{6}$$

It can be shown by using measure theory that the order of expectation and summation can be reversed in (6). Thus

$$ES_T = x + \sum_{j=1}^{\infty} E[X_j(1 - 1_{\{T<j\}})]. \tag{7}$$

In determining whether or not $T < j$, it suffices to look at the random variables $X_1, X_2, \ldots, X_{j-1}$. It follows that the random variables X_j and $(1 - 1_{\{T<j\}})$ are independent. Consequently

$$\begin{aligned} E[X_j(1 - 1_{\{T<j\}})] &= EX_jE(1 - 1_{\{T<j\}}) \\ &= \mu(1 - P(T < j)) \\ &= \mu P(T \geq j). \end{aligned}$$

We now conclude from (6) and Theorem 5 of Chapter 4 that

$$\begin{aligned} ES_T &= x + \mu \sum_{j=1}^{\infty} P(T \geq j) \\ &= x + \mu ET, \end{aligned}$$

which completes the proof of Wald's identity.

If the X_n's have mean $\mu = 0$ and finite variance σ^2, there is a second identity due to Wald, namely

$$\operatorname{Var} S_T = \sigma^2 ET. \tag{8}$$

Since $ES_T = x$ by (4), Formula (8) is equivalent to

$$E(S_T - x)^2 = \sigma^2 ET. \tag{9}$$

We will now verify (9). By (5)

$$S_T - x = \sum_{j=1}^{\infty} X_j(1 - 1_{\{T<j\}}),$$

and hence

$$\begin{aligned} (S_T - x)^2 &= \sum_{j=1}^{\infty} X_j(1 - 1_{\{T<j\}}) \sum_{k=1}^{\infty} X_k(1 - 1_{\{T<k\}}) \\ &= \sum_{j=1}^{\infty} \sum_{k=1}^{\infty} X_j(1 - 1_{\{T<j\}})X_k(1 - 1_{\{T<k\}}). \end{aligned}$$

In taking expectations it is again permissible to interchange expectation and summation. We conclude that

$$(10) \qquad E(S_T - x)^2 = \sum_{j=1}^{\infty} \sum_{k=1}^{\infty} E[X_j(1 - 1_{\{T<j\}})X_k(1 - 1_{\{T<k\}})].$$

We will now evaluate the individual terms in this double sum. Consider first terms corresponding to values of j and k such that $j < k$. The random variable

$$X_j(1 - 1_{\{T<j\}})(1 - 1_{\{T<k\}})$$

depends only on $X_1, X_2, \ldots, X_{k-1}$, and hence is independent of the random variable X_k. Since $EX_k = \mu = 0$, we see that

$$E[X_j(1 - 1_{\{T<j\}})X_k(1 - 1_{\{T<k\}})]$$
$$= E[X_j(1 - 1_{\{T<j\}})(1 - 1_{\{T<k\}})]EX_k = 0.$$

Similarly the terms in the right side of (10) vanish when $j > k$. When $j = k$ we obtain

$$E[X_j^2(1 - 1_{\{T<j\}})^2].$$

The random variable $(1 - 1_{\{T<j\}})$ depends only on $X_1, X_2, \ldots, X_{j-1}$, and hence is independent of X_j. Since this random variable takes on only the values 0 and 1, we see that

$$(1 - 1_{\{T<j\}})^2 = 1 - 1_{\{T<j\}}.$$

Thus

$$\begin{aligned} E[X_j^2(1 - 1_{\{T<j\}})^2] &= E[X_j^2(1 - 1_{\{T<j\}})] \\ &= EX_j^2 E(1 - 1_{\{T<j\}}) \\ &= \text{Var } X_j^2(1 - P(T < j)) \\ &= \sigma^2 P(T \geq j). \end{aligned}$$

We now conclude from (10) and Theorem 5 of Chapter 4 that

$$E(S_T - x)^2 = \sigma^2 \sum_{j=1}^{\infty} P(T \geq j) = \sigma^2 ET,$$

which proves (9) and hence also (8).

9.2. Simple random walks

We will assume throughout this section that $a \leq x \leq b$, $a < b$, and a, b, and x are integers. The two identities of the previous section are most easily applied if it is known that

$$(11) \qquad P(S_T = a \text{ or } b) = P(S_T = a) + P(S_T = b) = 1.$$

This will certainly be the case if

(12) $$P(X_k = -1, 0, \text{ or } 1) = 1,$$

in which case the random walk $S_0, S_1, \ldots$ is called a *simple random walk*. The main property that distinguishes simple random walks from other random walks is that they do not "jump over" integer points. Set

$$p = P\{X_k = 1\},$$
$$q = P\{X_k = -1\},$$
$$r = P\{X_k = 0\}.$$

Then $p \geq 0$, $q \geq 0$, $r \geq 0$, and $p + q + r = 1$. The assumption (2) states that $r < 1$. It follows from (11) that

(13) $$\begin{aligned} ES_T &= aP(S_T = a) + bP(S_T = b) \\ &= a(1 - P(S_T = b)) + bP(S_T = b). \end{aligned}$$

For simple random walks we can solve explicitly for $P(S_T = a)$, $P(S_T = b)$, ES_T, and ET. Consider first, the case $p = q$. Then $\mu = 0$ and Wald's identity (4) becomes $ES_T = x$. Thus by (13)

$$x = a(1 - P(S_T = b)) + bP(S_T = b).$$

It follows immediately that

(14) $$P(S_T = b) = \frac{x - a}{b - a}$$

and

(15) $$P(S_T = a) = \frac{b - x}{b - a}.$$

Wald's identity does not give any information about ET when $\mu = 0$. The identity (8) is applicable in this case, however, and we find that $\sigma^2 = p + q = 1 - r$ and

$$\operatorname{Var} S_T = \sigma^2 ET = (1 - r)ET.$$

Now

$$\begin{aligned} \operatorname{Var} S_T &= ES_T^2 - (ES_T)^2 \\ &= b^2 P(S_T = b) + a^2 P(S_T = a) - x^2 \\ &= \frac{b^2(x - a) + a^2(b - x)}{b - a} - x^2 \\ &= (ax + bx - ab) - x^2 \\ &= (x - a)(b - x). \end{aligned}$$

Thus if $p = q$,

(16) $$ET = \frac{(x - a)(b - x)}{1 - r}.$$

If $r = 0$ and $p = q = 1/2$,

$$ET = (x - a)(b - x). \tag{17}$$

Example 1. Two players having respective initial capitals of \$5 and \$10 agree to make a series of \$1 bets until one of them goes broke. Assume the outcomes of the bets are independent and both players have probability 1/2 of winning any given bet. Find the probability that the player with the initial capital of \$10 goes broke. Find the expected number of bets.

The problem fits into our scheme with S_n denoting the capital of the less wealthy player after n bets if we choose $p = q = 1/2$, $x = 5$, $a = 0$, and $b = 15$. The answer to the first part is given by

$$P(S_T = b) = \frac{5 - 0}{15 - 0} = \frac{1}{3}.$$

The answer to the second part is

$$ET = (5 - 0)(15 - 5) = 50.$$

Suppose next that $p \neq q$. To avoid trivial exceptions we will also assume that $p > 0$ and $q > 0$. Wald's identity cannot be used to find $P(S_T = b)$ if $p \neq q$; therefore another approach is required.

Define $f(x)$ for x an integer in $[a, b]$ by letting $f(x)$ denote the probability that $S_T = b$ when $S_0 = x$. We observe first that f satisfies the difference equation

$$f(x) = pf(x + 1) + qf(x - 1) + rf(x), \qquad a < x < b. \tag{18}$$

This is true because

$$\begin{aligned} f(x) &= P(S_T = b) \\ &= p \cdot P(S_T = b \mid X_1 = 1) + q \cdot P(S_T = b \mid X_1 = -1) \\ &\quad + r \cdot P(S_T = b \mid X_1 = 0) \end{aligned}$$

and

$$P(S_T = b \mid X_1 = i) = f(x + i), \qquad i = 1, -1, 0.$$

In addition to (18) we have the obvious boundary conditions

$$f(a) = 0 \quad \text{and} \quad f(b) = 1. \tag{19}$$

From (18) and $(1 - r) = p + q$, we see that

$$f(x + 1) - f(x) = \frac{q}{p}(f(x) - f(x - 1)), \qquad a < x < b. \tag{20}$$

Set $c = f(a + 1) = f(a + 1) - f(a)$. Then (20) implies that

$$f(x + 1) - f(x) = c\left(\frac{q}{p}\right)^{x-a}, \qquad a \le x < b.$$

Using the formula for the sum of a geometrical progression we obtain

$$f(x) = f(x) - f(a) = \sum_{y=a}^{x-1} (f(y+1) - f(y))$$

$$= \sum_{y=a}^{x-1} c\left(\frac{q}{p}\right)^{y-a}$$

$$= c\,\frac{1 - (q/p)^{x-a}}{1 - (q/p)}, \qquad a \le x \le b.$$

From the special case $f(b) = 1$, we now have that

$$c = \frac{1 - (q/p)}{1 - (q/p)^{b-a}}.$$

We substitute this in our expression for $f(x)$ and obtain

$$f(x) = \frac{1 - (q/p)^{x-a}}{1 - (q/p)^{b-a}}.$$

We have shown, in other words, that if $p \ne q$ and $p > 0$, then

$$(21) \qquad P(S_T = b) = \frac{1 - (q/p)^{x-a}}{1 - (q/p)^{b-a}}, \qquad a \le x \le b.$$

It follows immediately from (21) that under the same conditions

$$(22) \qquad P(S_T = a) = \frac{(q/p)^{x-a} - (q/p)^{b-a}}{1 - (q/p)^{b-a}}, \qquad a \le x \le b.$$

From (13) and (21)

$$(23) \qquad ES_T = (b - a)\,\frac{1 - (q/p)^{x-a}}{1 - (q/p)^{b-a}} + a.$$

Since $\mu = p - q$, it now follows from Wald's identity that

$$(24) \qquad ET = \left(\frac{b-a}{p-q}\right)\frac{1 - (q/p)^{x-a}}{1 - (q/p)^{b-a}} - \frac{x-a}{p-q}, \qquad a \le x \le b.$$

Example 2. Let us modify the previous example by supposing that the wealthier player because of his greater skill has probability .6 of winning any given bet, with the other player having probability .4 of winning the bet. Find the probability that the wealthier player goes broke, the expected gain to that player, and the expected number of bets.

Here we take $p = .4$ and $q = .6$. The probability that the wealthier player goes broke is

$$P(S_T = 15) = \frac{1 - (.6/.4)^5}{1 - (.6/.4)^{15}} = .0151.$$

In order to find the expected gain to the wealthier player we first note that the expected capital of the poorer player after they stop playing is

$$ES_T = 15P(S_T = 15) = 15(.0151) = .23.$$

Thus the expected gain to the wealthier player or the expected loss to the other player is \$5.00 − \$.23 = \$4.77, a good percentage of the poorer player's initial capital. The expected number of bets is

$$ET = \frac{ES_T - x}{\mu} = \frac{-4.77}{-.2} \approx 24.$$

Let $b \to \infty$ in (21). It can be shown that the left side of (21) converges to $P(S_n > a \text{ for all } n \geq 0)$. If $q < p$, the right side of (21) converges to $1 - (q/p)^{x-a}$. If $q > p$, the right side of (21) converges to 0. If $q = p$, the right side of (14) converges to 0. Thus for $a \leq x = S_0$,

$$(25) \qquad P(S_n > a \text{ for all } n \geq 0) = \begin{cases} 1 - \left(\dfrac{q}{p}\right)^{x-a} & \text{for } q < p, \\ 0 & \text{for } q \geq p. \end{cases}$$

Similarly for $b \geq x = S_0$,

$$P(S_n < b \text{ for all } n \geq 0) = \begin{cases} 1 - \left(\dfrac{p}{q}\right)^{b-x} & \text{for } p < q, \\ 0 & \text{for } p \geq q. \end{cases}$$

Example 3. A gambling house has a capital of one hundred thousand dollars. An infinitely rich gambler tries to break the house. He is allowed to try, and decides to bet \$1000 at a time. If the gambler has the probabilities .49 of winning each bet and .51 of losing, what is the probability he will ever break the house?

Let S_n denote the capital of the house (in multiples of \$1000) after n games. Then $p = .51$, $q = .49$, $x = 100$, and $a = 0$. By (25) the probability that the house will go broke is

$$1 - P(S_n > 0 \text{ for all } n \geq 0) = \left(\frac{q}{p}\right)^{x-a} = \left(\frac{.49}{.51}\right)^{100} = .018.$$

Let A be a subset of the integers (in applications, A will have 0, 1, or 2 points). For $x \notin A$ and $y \notin A$, let $P_A(x, y)$ denote the probability that a simple random walk beginning at x will hit y at some positive time before hitting A. For $x \in A$ or $y \in A$ set $P_A(x, y) = 0$. These probabilities can be computed in terms of the formulas of this section.

Example 4. Suppose $p = q$. Find $P_{\{a,b\}}(y, y)$, where $a < y < b$.

After one step, the random walk is at $y - 1$, y, or $y + 1$ with respective probabilities p, $1 - 2p$, and p. From $y - 1$, the probability of returning to y before a is $(y - a - 1)/(y - a)$. From $y + 1$, the probability of returning to y before b is $(b - y - 1)/(b - y)$. Thus the probability of returning to y before hitting a or b is given by

$$P_{\{a,b\}}(y, y) = p\,\frac{y - a - 1}{y - a} + 1 - 2p + p\,\frac{b - y - 1}{b - y}$$

or

$$P_{\{a,b\}}(y, y) = 1 - \frac{p(b - a)}{(y - a)(b - y)}. \tag{26}$$

For $x \notin A$ and $y \notin A$, let $G_A(x, y)$ denote the expected number of visits to y (for positive values of n) before hitting A for a random walk starting at x. Set $G_A(x, y) = 0$ if $x \in A$ or $y \in A$. It is not hard to show that the number of returns from y to y before hitting A has a geometric distribution with parameter $p = 1 - P_A(y, y)$. Thus by Example 3 of Chapter 4,

$$G_A(y, y) = \frac{P_A(y, y)}{1 - P_A(y, y)}. \tag{27}$$

If $x \neq y$, then

$$G_A(x, y) = P_A(x, y)(1 + G_A(y, y)). \tag{28}$$

For to have any positive number of visits to y before hitting A, we must first get to y before hitting A. This has probability $P_A(x, y)$. If we do get to y before hitting A, then the total number of visits to y before hitting A is 1 plus the total number of returns from y to y before hitting A. This explains (28). From (27) and (28) we have that

$$G_A(x, y) = \frac{P_A(x, y)}{1 - P_A(y, y)} \qquad \text{for all } x \text{ and } y. \tag{29}$$

Example 5. Let us return to the first example of this section. Find the expected number of times $n > 1$ that two players will be back to their initial capital before one of them goes broke.

We recall that in this example $p = q = 1/2$, $a = 0$, $x = 5$, and $b = 15$. The probability of returning back to the original capitals before one of the players goes broke is, by (26),

$$P_{\{0,15\}}(5, 5) = 1 - \frac{(1/2)(15)}{5 \cdot 10} = .85.$$

Thus by (27) the expected number of times both players will be back to their initial capital before one of them goes broke is

$$G_{\{0,15\}}(5, 5) = \frac{P_{\{0,15\}}(5, 5)}{1 - P_{\{0,15\}}(5, 5)}$$

$$= \frac{.85}{.15} = 5.67.$$

9.3. Construction of a Poisson process

In the remaining sections of this chapter we will consider a probabilistic model for the random distribution of particles in space or events in time. Such models find applications in a variety of different fields. As an example, suppose we have a piece of radioactive material and let an experiment consist of observing the times when disintegrations occur. Suppose the experiment starts at time 0 and let D_m denote the time of the mth disintegration. As discussed in Example 2 of Chapter 1, the laws of physics tell us that the times D_m must be considered random variables. The collection of points $\{D_1, D_2, \ldots\}$ can be considered as a random countable subset of $[0, \infty)$.

As another illustration of essentially the same phenomena, consider calls coming into a telephone exchange. Let D_m denote the time when the mth call begins. There is no known way of predicting the D_m exactly, but they can usefully be treated as random variables.

Consider an experiment of the following type. A swab is dipped into a test tube containing a suspension of bacteria and is then smeared uniformly across the surface of a disk containing nutrients upon which the bacteria can multiply. After a few days, wherever a bacterium was dropped there appears a visible colony of bacteria. The locations of these spots as well as their total number are random. This situation is illustrated in Figure 2.

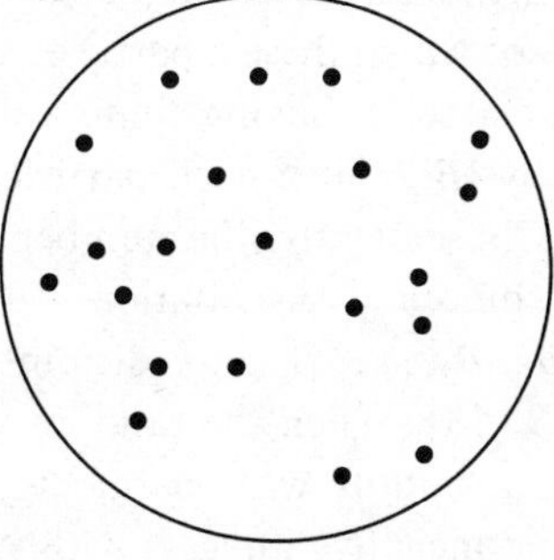

Figure 2

The location of the spots can be viewed as a random subset of points of the disk.

In the examples of radioactive particles and bacteria, we were led to consider a random collection of points in a certain subset S of Euclidean space. In these examples both the location of the "particles" and their total number is random. Associated with such a random collection are various random variables such as the total number N of particles in S, the number of particles N_B in a specified subset $B \subset S$, and the distance D_m from a specified point in S to the mth closest particle. In Figure 3 the

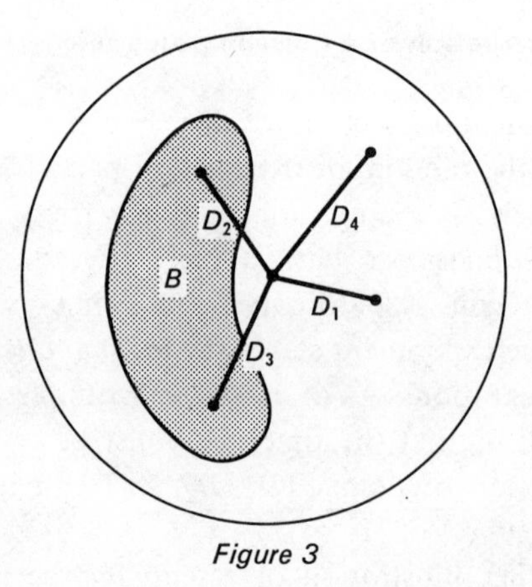

Figure 3

particles are denoted by dots. In that figure $N = N_S = 4$ and $N_B = 2$. It is the distributions and joint distributions of random variables of this type that we will be studying in the rest of this chapter.

Naturally, there are many different mathematical models that consider the random distribution of particles. We will consider one of the most elementary and most important such models, called the Poisson process. Such a process is closely related to the uniform distribution of particles which we will discuss first.

Consider then a system in which the total number of particles n is fixed, but for which the locations of the particles in S are random. The model we want is one for which these n particles are independently and uniformly distributed over a set S having finite volume. Denote the volume of a subset B of S by $|B|$. Then each particle has probability $p = |B|/|S|$ of landing in B. Consequently, the number of particles N_B that land in the set B has the binomial distribution with parameters n and p. More generally, let $B_1, B_2, \ldots, B_k$ be k disjoint subsets of S whose union is S and let $p_j = |B_j|/|S|$. Then the random variables $N_{B_1}, \ldots, N_{B_k}$ have the multinomial distribution with parameters n and $p_1, \ldots, p_k$. Hence if $n_1, \ldots, n_k$ are nonnegative integers with sum n,

$$P(N_{B_1} = n_1, \ldots, N_{B_k} = n_k) = \frac{n!}{(n_1!) \cdots (n_k!)} p_1^{n_1} \cdots p_k^{n_k}$$

$$= \frac{n!}{|S|^n} \prod_{j=1}^{k} \frac{|B_j|^{n_j}}{n_j!}.$$

The Poisson process on S is a modification of the above. We suppose now that the total number of particles $N = N_S$ in S is a random variable having a Poisson distribution with parameter $\lambda|S|$. Moreover, we assume that given $N = n$, the n particles are independently and uniformly distributed over S. Let $B_1, \ldots, B_k$ be as before. Then our assumptions are that if $n_1, \ldots, n_k$ are k nonnegative integers with sum n, then

$$P(N_{B_1} = n_1, \ldots, N_{B_k} = n_k \mid N = n) = \frac{n!}{|S|^n} \prod_{j=1}^{k} \frac{|B_j|^{n_j}}{n_j!}.$$

Hence

$$(30) \quad P(N = n, N_{B_1} = n_1, \ldots, N_{B_k} = n_k) = P(N = n) \frac{n!}{|S|^n} \prod_{j=1}^{k} \frac{|B_j|^{n_j}}{n_j!}$$

$$= \frac{\lambda^n |S|^n}{n!} e^{-\lambda|S|} \frac{n!}{|S|^n} \prod_{j=1}^{k} \frac{|B_j|^{n_j}}{n_j!}$$

$$= \lambda^n e^{-\lambda|S|} \prod_{j=1}^{k} \frac{|B_j|^{n_j}}{n_j!}.$$

Since the sets B_j are disjoint with union S,

$$|S| = |B_1| + \cdots + |B_k|,$$

and thus we can write the right-hand side of (30) as

$$\prod_{j=1}^{k} \frac{(\lambda|B_j|)^{n_j}}{n_j!} e^{-\lambda|B_j|}.$$

Now the event $\{N = n, N_{B_1} = n_1, \ldots, N_{B_k} = n_k\}$ is the same as the event $\{N_{B_1} = n_1, \ldots, N_{B_k} = n_k\}$, because $n = n_1 + \cdots + n_k$ and $N = N_{B_1} + \cdots + N_{B_k}$. Thus

$$P(N = n, N_{B_1} = n_1, \ldots, N_{B_k} = n_k) = P(N_{B_1} = n_1, \ldots, N_{B_k} = n_k).$$

We have therefore proved the important fact:

$$(31) \qquad P(N_{B_1} = n_1, \ldots, N_{B_k} = n_k) = \prod_{j=1}^{k} \frac{(\lambda|B_j|)^{n_j}}{n_j!} e^{-\lambda|B_j|}.$$

In other words, *the random variables $N_{B_1}, \ldots, N_{B_k}$ are independent and Poisson distributed with parameters $\lambda|B_j|$ respectively.*

It is not too surprising that the random variables N_{B_j}, $1 \le j \le k$, are Poisson distributed; but it is surprising that they are mutually independent because in the case where the number of particles is fixed the corresponding quantities are dependent. It is this independence property that makes the Poisson process easy to work with in applications.

In the preceding models the total number of particles, whether random or fixed, was always finite, and they were distributed over sets having finite total volume. For some purposes, however, it is theoretically simpler to consider an infinite number of particles distributed over a set having infinite volume. Thus we might want to distribute particles over all of R^r or over $[0, \infty)$, etc. To cover such cases, we require only a slight extension of the previous model.

The basic assumption of a Poisson process on a set S having finite or infinite volume is that if $B_1, B_2, \ldots, B_k$ are disjoint subsets of S each having finite volume, then the random variables $N_{B_1}, \ldots, N_{B_k}$ are independent and Poisson distributed with parameters $\lambda|B_1|, \ldots, \lambda|B_k|$ respectively. The constant λ is called the parameter of the process.

Let B be a subset of S having finite volume. As a consequence of the definition of a Poisson process, it follows that if $B_1, B_2, \ldots, B_k$ are disjoint subsets of B whose union is B, and $n_1, n_2, \ldots, n_k$ are nonnegative integers with sum n, then

$$(32) \qquad P(N_{B_1} = n_1, \ldots, N_{B_k} = n_k \mid N_B = n) = \frac{n!}{|B|^n} \prod_{j=1}^{k} \frac{|B_j|^{n_j}}{n_j!}.$$

To verify (32) note that the desired probability is just

$$\frac{P(N_{B_1} = n_1, \ldots, N_{B_k} = n_k)}{P(N_B = n)} = \frac{\prod_{j=1}^{k} e^{-\lambda|B_j|} (\lambda|B_j|)^{n_j}/n_j!}{e^{-\lambda|B|} (\lambda|B|)^n/n!},$$

which reduces to the right-hand side of (32).

Another way of looking at Equation (32) is as follows: Given that there are n particles in B, the joint distribution of the random variables $N_{B_1}, \ldots, N_{B_k}$ is the same as that obtained by distributing n particles independently and uniformly over B. This fact is very useful for solving some problems in which the Poisson process acts as an input to a more complicated system. We will not pursue this aspect of a Poisson process any further in the text. (See, however, Exercises 21 and 31 for simple illustrations of its use.)

9.4. Distance to particles

Suppose we have a Poisson process on a subset S of Euclidean space. If S has finite volume, then the number N of particles in S is finite. Let $D_1 \le D_2 \le \cdots \le D_N$ denote the distance to these particles from the origin arranged in nondecreasing order. If S has infinite volume, then the number of particles in S is infinite, and we let $D_1 \le D_2 \le \cdots \le D_m \cdots$ denote the distances to the particles arranged in nondecreasing order. Such an arrangement is possible because for any positive number r only finitely many particles are at distance less than r from the origin. In this section we will compute the distribution of D_m for various choices of the set S.

We first give an example where these distances enter in a natural way. Suppose that stars in a certain set S of 3-dimensional Euclidean space are distributed according to a Poisson process on S with parameter λ. Suppose further that these stars are equally bright. The amount of light reaching the origin from a star is proportional to the inverse square of the distance from the origin to the star. Thus the amount of light received from a star of distance r from the origin is K/r^2 for some positive constant K. The amount of light received from the nearest (and hence apparently brightest) star is K/D_1^2. The total amount of light is

$$\sum_m \frac{K}{D_m^2}.$$

By advanced probability techniques it can be shown that if S is all of three-dimensional space, then the sum of the above series is infinite with probability one. This fact has interesting implications in cosmology.

We will now compute the distribution of D_m, assuming for simplicity that S has infinite volume. Let $S_r = S \cap \{x : |x| \le r\}$ (that is, let S_r be the set of points in S at distance at most r from the origin), and let $\varphi(r)$ denote the volume of S_r. The number N_{S_r} of particles in S_r has a Poisson distribution with parameter $\lambda\varphi(r)$. The event $\{D_m \le r\}$ is the same as the event $\{N_{S_r} \ge m\}$. Thus by Equations (39) and (40) of Chapter 5,

$$(33) \qquad P(D_m \le r) = P(N_{S_r} \ge m) = \int_0^{\lambda\varphi(r)} \frac{t^{m-1}e^{-t}}{(m-1)!}\,dt.$$

It follows from (33) that if $\varphi(r)$ is differentiable, then D_m has a density function f_m given by

$$(34) \qquad f_m(r) = \frac{\lambda^m \varphi(r)^{m-1}\varphi'(r)e^{-\lambda\varphi(r)}}{(m-1)!}, \qquad r > 0.$$

If $\varphi(r)$ is strictly increasing it has a continuous inverse function $\varphi^{-1}(r)$. It follows from (33) that the random variable $\varphi(D_m)$ has the gamma distribution $\Gamma(m, \lambda)$.

In several important cases $\varphi(r)$ is of the form $\varphi(r) = cr^d$, where c is a positive numerical constant (this is true, for example, if $S = R^d$ or if $S = [0, \infty)$). In this case (34) becomes

$$(35) \qquad f_m(r) = \frac{d(c\lambda)^m}{(m-1)!} r^{md-1}e^{-c\lambda r^d}, \qquad r > 0.$$

Various special cases of this formula will be considered in the exercises.

We will use Formula (35) to compute the moments ED_m^j in these cases. Thus

$$ED_m^j = \int_0^\infty r^j f_m(r)\,dr = \int_0^\infty \frac{d(c\lambda)^m}{(m-1)!} r^{md+j-1}e^{-c\lambda r^d}\,dr.$$

To compute this, we make the change of variable $s = \lambda c r^d$ and integrate to obtain

$$ED_m^j = \frac{(\lambda c)^{-j/d}\Gamma(m + (j/d))}{(m - 1)!}.$$

9.5. Waiting times

So far we have been visualizing a Poisson process as a model for the distribution of particles in space. If we think of the set $[0, \infty)$ as the time axis, then we may consider a Poisson process on $[0, \infty)$ as the distribution of times at which certain events occur. In the beginning of Section 3 we mentioned some examples of using a Poisson process on $[0, \infty)$ in this manner.

In the switch from thinking of a Poisson process on $[0, \infty)$ as a distribution of particles in the set $[0, \infty)$ to thinking of the Poisson process as the distribution of events in time, a new set of terms is introduced. Instead of speaking of "particles," we now speak of "events", and the distance D_m to the mth particle now becomes the time when the mth event takes place.

Let $N(t) = N_{[0,t]}$ denote the number of events that occur during the time span $[0, t]$. Then $N(t)$ is a Poisson distributed random variable with parameter λt. If $0 \le s \le t$, then $N(t) - N(s)$ represents the number of events taking place during the time span $(s, t]$, and it has a Poisson distribution with parameter $\lambda(t - s)$. More generally, if $0 \le t_1 \le t_2 \le \cdots \le t_n$, then $N(t_1), N(t_2) - N(t_1), \ldots, N(t_n) - N(t_{n-1})$ are independent random variables having Poisson distributions with parameters

$$\lambda t_1, \lambda(t_2 - t_1), \ldots, \lambda(t_n - t_{n-1})$$

respectively. These facts are immediate from our definition of the Poisson process and its translation into the time language.

As mentioned above, D_m is the time of the mth event. From our results in Section 9.4 we know that D_m has the gamma distribution $\Gamma(m, \lambda)$. In particular, D_1 is exponentially distributed with parameter λ. Recall from Chapter 6 that the sum of m independent, identically distributed exponential variables has the gamma distribution $\Gamma(m, \lambda)$. Define random variables $W_1, W_2, \ldots, W_n, \ldots$ as follows: $W_1 = D_1$, $W_n = D_n - D_{n-1}$, $n \ge 2$. Then clearly $D_m = W_1 + \cdots + W_m$. The discussion just given makes it plausible that $W_1, W_2, \ldots, W_m$ are independent exponentially distributed random variables with the common parameter λ. This is in fact true and is a very interesting and useful property of the Poisson process on $[0, \infty)$. The random variable W_m is, of course, nothing more than the time between the $(m - 1)$st event and the mth, so that the times $W_1, W_2, \ldots$ are the *waiting times* between successive events in a Poisson process.

Theorem 1 *Let* $W_1, W_2, \ldots, W_n, \ldots$ *be the waiting times between successive events in a Poisson process on* $[0, \infty)$ *with parameter* λ. *Then* W_n, $n \geq 1$, *are mutually independent exponentially distributed random variables with the common mean* λ^{-1}.

Proof. Let f_n be the n-dimensional density given by

$$f_n(t_1, \ldots, t_n) = \begin{cases} \lambda^n e^{-\lambda t_n} & \text{for } 0 \leq t_1 \leq t_2 \leq \cdots \leq t_n, \\ 0, & \text{elsewhere.} \end{cases}$$

From Example 13 of Chapter 6, we see that the theorem is true if and only if the random variables $D_1, \ldots, D_n$ have joint density f_n. This is true for $n = 1$, since D_1 is exponentially distributed with parameter λ.

A rigorous proof in general is more complicated. Before giving this proof we will first give a heuristic way of seeing this fact.

Let $0 = t_0 < t_1 < \cdots < t_n$ and choose $h > 0$ so small that $t_{i-1} + h \leq t_i$, $i = 1, 2, \ldots, n$. Then (see Figure 4)

$$\begin{aligned} (36) \qquad & P(t_i < D_i \leq t_i + h, 1 \leq i \leq n) \\ &= P(N(t_1) = 0, N(t_1 + h) - N(t_1) = 1, \ldots, \\ &\qquad N(t_n) - N(t_{n-1} + h) = 0, N(t_n + h) - N(t_n) \geq 1) \\ &= e^{-\lambda t_1}(\lambda h)e^{-\lambda h} \cdots e^{-\lambda(t_n - t_{n-1} - h)}[1 - e^{-\lambda h}] \\ &= \lambda^{n-1} h^{n-1} e^{-\lambda t_n}(1 - e^{-\lambda h}). \end{aligned}$$

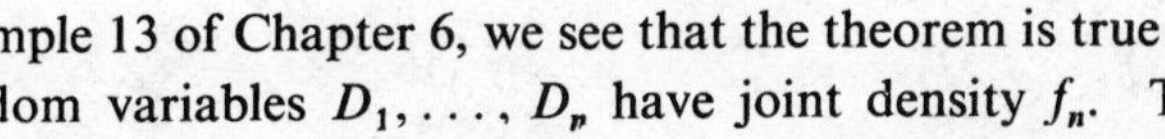

0 1 0 1 0

t_1 t_1+h t_2 t_2+h t_3

Figure 4

If we knew that the random variables $D_1, D_2, \ldots, D_n$ had a joint density g_n that was continuous at the point $(t_1, t_2, \ldots, t_n)$, we could conclude that

$$P(t_i < D_i \leq t_i + h \text{ for } 1 \leq i \leq n) = g_n(t_1, \ldots, t_n)h^n + e(h),$$

where $e(h)$ is some function of h such that

$$\lim_{h \downarrow 0} \frac{e(h)}{h^n} = 0.$$

It would then follow from (36) that

$$\begin{aligned} g(t_1, \ldots, t_n) &= \lim_{h \downarrow 0} h^{-n} P(t_i < D_i \leq t_i + h) \\ &= \lim_{h \downarrow 0} \lambda^{n-1} e^{-\lambda t_n} \frac{1 - e^{-\lambda h}}{h} \\ &= \lambda^n e^{-\lambda t_n} \end{aligned}$$

as desired.

We will now give an elementary but rigorous proof of Theorem 1. Although this proof is not difficult, it is rather long, and the reader may wish to omit it.

Let F_n be the distribution function having density f_n. It follows at once from the definition of f_n that for $n \geq 2$,

$$f_n(s_1, \ldots, s_n) = f_1(s_1)f_{n-1}(s_2 - s_1, \ldots, s_n - s_1).$$

Integrating both sides over the set $s_1 \leq t_1, \ldots, s_n \leq t_n$, we see that

$$(37) \qquad F_n(t_1, \ldots, t_n) = \int_0^{t_1} f_1(s_1)F_{n-1}(t_2 - s_1, \ldots, t_n - s_1)\, ds_1.$$

Let G_n denote the joint distribution of $D_1, D_2, \ldots, D_n$. From Example 10 of Chapter 6 we see that the theorem is true if and only if the random variables $D_1, \ldots, D_n$ have joint density f_n, and hence joint distribution function F_n. Consequently, to prove our theorem we must show that $F_n = G_n$.

Now F_1 is just the exponential distribution with parameter λ. As pointed out before, G_1, the distribution of D_1, is also exponential with parameter λ. Thus $F_1 = G_1$.

Suppose we can show that for $n \geq 2$,

$$(38) \qquad G_n(t_1, \ldots, t_n) = \int_0^{t_1} f_1(s_1)G_{n-1}(t_2 - s_1, \ldots, t_n - s_1)\, ds_1.$$

Then, as $G_1 = F_1$ it would follow from (37) and (38) that $G_2 = F_2$. Using the fact that $G_2 = F_2$, another application of (37) and (38) would show $G_3 = F_3$, etc. In other words, if we knew that (38) held, then our theorem would follow from (37) and (38) by induction. In establishing (38) we can assume that $0 \leq t_1 \leq \cdots \leq t_n$, since otherwise both sides of (38) are zero.

To start on the proof, observe that the event $\{D_i \leq t_i\}$ is the same as the event $\{N(t_i) \geq i\}$ and thus

$$\begin{aligned}\{D_i \leq t_i, 1 \leq i \leq n\} &= \bigcap_{i=1}^{n} \{D_i \leq t_i\} \\ &= \bigcap_{i=1}^{n} \{N(t_i) \geq i\} \\ &= \{N(t_i) \geq i, 1 \leq i \leq n\}.\end{aligned}$$

We can therefore write

$$G_n(t_1, t_2, \ldots, t_n) = P(N(t_i) \geq i, 1 \leq i \leq n).$$

Consequently (38) is the same as

$$(39) \qquad P(N(t_i) \geq i, 1 \leq i \leq n) = \int_0^{t_1} f_1(s_1)P(N(t_i - s_1) \geq i - 1, 2 \leq i \leq n)\, ds_1.$$

To establish (39), note first that for any $k \geq 1$

$$\frac{(\lambda t)^k}{k!} e^{-\lambda t} = \int_0^t \lambda e^{-\lambda s} \frac{[\lambda(t-s)]^{k-1}}{(k-1)!} e^{-\lambda(t-s)}\, ds. \tag{40}$$

Indeed,

$$\begin{aligned} \int_0^t \lambda e^{-\lambda s} \frac{[\lambda(t-s)]^{k-1}}{(k-1)!} e^{-\lambda(t-s)}\, ds &= \frac{e^{-\lambda t}\lambda^k}{(k-1)!} \int_0^t (t-s)^{k-1}\, ds \\ &= \frac{e^{-\lambda t}\lambda^k}{(k-1)!} \int_0^t s^{k-1}\, ds = \frac{(\lambda t)^k}{k!} e^{-\lambda t}. \end{aligned}$$

Now let $0 \leq t_1 \leq t_2 \leq \cdots \leq t_n$ and let $1 \leq k_1 \leq k_2 \leq \cdots \leq k_n$. Next we claim that

$$\begin{aligned} &P(N(t_1) = k_1, \ldots, N(t_n) = k_n) \\ &\quad = \int_0^{t_1} \lambda e^{-\lambda s} P(N(t_1 - s) = k_1 - 1, \ldots, N(t_n - s) = k_n - 1)\, ds. \end{aligned} \tag{41}$$

To see this observe that by (40)

$$\begin{aligned} &P(N(t_1) = k_1, \ldots, N(t_n) = k_n) \\ &\quad = P(N(t_1) = k_1) \prod_{j=2}^{n} P(N(t_j) - N(t_{j-1}) = k_j - k_{j-1}) \\ &\quad = \int_0^{t_1} \lambda e^{-\lambda s} \frac{[\lambda(t_1 - s)]^{k_1 - 1} e^{-\lambda(t_1 - s)}\, ds}{(k_1 - 1)!} \\ &\qquad \times \prod_{j=2}^{n} \frac{[\lambda(t_j - t_{j-1})]^{k_j - k_{j-1}} e^{-\lambda(t_j - t_{j-1})}}{(k_j - k_{j-1})!}. \end{aligned} \tag{42}$$

On the other hand,

$$\begin{aligned} &\int_0^{t_1} \lambda e^{-\lambda s} P(N(t_1 - s) = k_1 - 1, \ldots, N(t_n - s) = k_n - 1)\, ds \\ &\quad = \int_0^{t_1} \lambda e^{-\lambda s} P(N(t_1 - s) = k_1 - 1) \\ &\qquad \times \prod_{j=2}^{n} P(N(t_j - s) - N(t_{j-1} - s) = k_j - k_{j-1})\, ds \\ &\quad = \int_0^{t_1} \lambda e^{-\lambda s} \frac{[\lambda(t_1 - s)]^{k_1 - 1} e^{-\lambda(t_1 - s)}}{(k_1 - 1)!} \\ &\qquad \times \prod_{j=2}^{n} \frac{[\lambda(t_j - t_{j-1})]^{k_j - k_{j-1}} e^{-\lambda(t_j - t_{j-1})}}{(k_j - k_{j-1})!}\, ds. \end{aligned} \tag{43}$$

Comparing the right-hand side of (42) with that of (43), we see that (41) holds. The desired equality (39) now follows from (41) by summing both sides of (41) over all values of $k_1, \ldots, k_n$ such that $k_1 \leq k_2 \leq \cdots \leq k_n$ and $k_1 \geq 1, k_2 \geq 2, \ldots, k_n \geq n$. ∎

Exercises

1 Let S_n be a random walk with $\mu = 0$ and $S_0 = x$. Suppose that

$$P(a - c \leq S_T \leq b + d) = 1,$$

where $a < x < b$, $c \geq 0$, and $d \geq 0$.
(a) Show that

$$(a - c)P(S_T \leq a) + bP(S_T \geq b) \leq x \leq aP(S_T \leq a) + (b + d)P(S_T \geq b).$$

(b) Show that

$$\frac{x - a}{b - a + d} \leq P(S_T \geq b) \leq \frac{x - a + c}{b - a + c}.$$

2 A gambler makes a series of bets of \$1. He decides to quit betting as soon as his net winnings reach \$25 or his net losses reach \$50. Suppose the probabilities of his winning and losing each bet are both equal to 1/2.
(a) Find the probability that when he quits he will have lost \$50.
(b) Find his expected loss.
(c) Find the expected number of bets he will make before quitting.

3 Suppose the gambler described in Exercise 2 is playing roulette and his true probabilities of winning and losing each bet are 9/19 and 10/19 respectively. Solve (a), (b), and (c) of Exercise 2 using the true probabilities.

4 A gambler makes a series of bets with probability p of winning and probability $q > p$ of losing each bet. He decides to play until he has either won M_1 dollars or lost M_2 dollars, where M_1 and M_2 are positive integers. He has a choice of betting 1 dollar at a time or of betting 1/2 dollar at a time. Show that he is more likely to win M_1 dollars before losing M_2 dollars if he bets 1 dollar at a time than if he bets 1/2 dollar at a time. What generalization of this result seems plausible?

5 Derive (14) by solving the appropriate difference equation.

6 Let S_n denote a simple random walk with $p = q = 1/2$ and let $a < b$. Find $P_{\{a,b\}}(x, y)$ and $G_{\{a,b\}}(x, y)$ for $a < x < b$ and $a < y < b$.

7 Let S_n denote a simple random walk with $p = q = 1/2$. Find $P_{\{0\}}(x, y)$ and $G_{\{0\}}(x, y)$ for $x > 0$ and $y > 0$.

8 Let S_n denote a simple random walk with $0 < q < p$. Find $P_{\varnothing}(x, y)$ and $G_{\varnothing}(x, y)$.

9 Let S_n denote a simple random walk with $0 < q < p$. Find $P_{\{0\}}(-1, y)$ and $G_{\{0\}}(-1, y)$ for $y < 0$.

10 Suppose points are distributed in 3-dimensional space according to a Poisson process with parameter $\lambda = 1$. Each point of the process is taken as the center of a sphere of radius r. Let X denote the number of spheres that contain the origin. Show that X is Poisson distributed with mean $(4/3)\pi r^3$.

11 A point is chosen at random in a circle with center at the origin and radius R. That point is taken as the center of a circle with radius X where X is a random variable having density f. Find the probability p that the random circle contains the origin.

12 Suppose n points are independently and uniformly chosen in the circle with center at the origin and radius R. Each point is taken as the center of a random circle whose radius has density f. Find, in terms of p of Exercise 11, the probability that exactly k circles contain the origin.

13 Find the answer to Exercise 12 if the n points are replaced with a random number N of points having a Poisson distribution with mean πR^2.

14 Suppose N balls are distributed at random into r boxes, where N is Poisson distributed with mean λ. Let Y denote the number of empty boxes. Show that Y is binomially distributed with parameters r and $p = e^{-\lambda/r}$. *Hint:* If X_i is the number of balls in box i, then $X_1, \ldots, X_r$ are independent Poisson distributed random variables each having mean λ/r.

15 Using the result of Exercise 14 we may easily derive the probability $p_k(r, n)$ that exactly k boxes are empty when n balls are distributed at random into r boxes. To do this first observe that

$$\begin{aligned} P(Y = k) &= \sum_{n=0}^{\infty} P(N = n)P(Y = k \mid N = n) \\ &= \sum_{n=0}^{\infty} e^{-\lambda} \frac{\lambda^n}{n!} p_k(r, n) \end{aligned}$$

and

$$P(Y = k) = \binom{r}{k} e^{-\lambda k/r}(1 - e^{-\lambda/r})^{r-k}.$$

So

$$\sum_{n=0}^{\infty} \frac{\lambda^n}{n!} p_k(r, n) = \binom{r}{k} e^{\lambda(r-k)/r}(1 - e^{-\lambda/r})^{r-k}.$$

Now equate coefficients of λ^n to rederive Equation (16) of Chapter 2.

16 Suppose the times of successive failures of a machine form a Poisson process on $[0, \infty)$ with parameter λ.

(a) What is the probability of at least one failure during the time period $(t, t + h)$, $h > 0$?

(b) What is the conditional probability of at least one failure by time $t + h$, given that there is no failure by time t?

17 Suppose we have a Poisson process on $[0, \infty)$ with parameter λ. Let Z_t denote the distance from t to the nearest particle to the right. Compute the distribution function of Z_t.

18 Suppose we have a Poisson process on $[0, \infty)$ with parameter λ. Let Y_t denote the distance from t to the nearest particle to the left. Take $Y_t = t$ if there are no particles to the left. Compute the distribution function of Y_t.

19 For Z_t and Y_t as in Exercises 17 and 18,
(a) show that Y_t and Z_t are independent,
(b) compute the distribution of $Z_t + Y_t$.

20 Particles arrive at a counter according to a Poisson process with parameter λ. Each particle gives rise to a pulse of unit duration. The particle is counted by the counter if and only if it arrives when no pulses are present. Find the probability that a particle is counted between time t and time $t + 1$. Assume $t \geq 1$.

21 Consider a Poisson process on $[0, \infty)$ with parameter λ and let T be a random variable independent of the process. Assume T has an exponential distribution with parameter ν. Let N_T denote the number of particles in the interval $[0, T]$. Compute the discrete density of N_T.

22 Do Exercise 21 if T has the uniform distribution on $[0, a]$, $a > 0$.

23 Consider two independent Poisson processes on $[0, \infty)$ having parameters λ_1 and λ_2 respectively. What is the probability that the first process has an event before the second process does?

24 Suppose n particles are distributed independently and uniformly on a disk of radius r. Let D_1 denote the distance from the center of the disk to the nearest particle. Compute the density of D_1.

25 For D_1 as in Exercise 24 compute the moments of D_1. *Hint:* Obtain a Beta integral by a change of variables.

26 Consider a Poisson process on R^r having parameter λ. For a set A having finite volume, let N_A denote the number of particles in A.
(a) Compute EN_A^2.
(b) If A and B are two sets having finite volume, compute $E(N_A N_B)$.

27 Let $A_1, A_2, \ldots, A_n$ be n disjoint sets having finite volume, and similarly let $B_1, B_2, \ldots, B_n$ be n disjoint sets having finite volume. For real numbers $\alpha_1, \ldots, \alpha_n$ and $\beta_1, \ldots, \beta_n$, set

$$f(x) = \sum_{i=1}^{n} \alpha_i 1_{A_i}(x)$$

and

$$g(x) = \sum_{i=1}^{n} \beta_i 1_{B_i}(x).$$

Show that for a Poisson process having parameter λ

$$E\left(\sum_{i=1}^{n} \alpha_i N_{A_i}\right)\left(\sum_{i=1}^{n} \beta_i N_{B_i}\right) = \lambda^2 \left(\int_{R^r} f(x)\, dx\right)\left(\int_{R^r} g(x)\, dx\right) + \lambda \int_{R^r} f(x)g(x)\, dx.$$

28 In Exercise 27 show that

$$\operatorname{Var}\left(\sum_{i=1}^{n} \alpha_i N_{A_i}\right) = \lambda \int_{R^r} f^2(x)\,dx.$$

29 Consider a Poisson process with parameter λ on R^3, and let D_m denote the distance from the origin to the mth nearest particle.
(a) Find the density of D_m.
(b) Find the density of D_m^3.

30 Suppose we have a Poisson process with parameter λ on the upper half-plane of R^2, i.e., the Poisson process is on the subset $S = \{(x, y): y > 0\}$ of R^2.
(a) What is the density of the distance D_m from 0 to the mth nearest particle?
(b) Find the first two moments of D_m.

31 Consider the following system. The times when particles arrive into the system constitute a Poisson process on $[0, \infty)$ with parameter λ. Each particle then lives for a certain length of time independent of the arrival times of the particles in the process and independent of the lives of the other particles. Suppose the lengths of life of the particles are exponentially distributed with common parameter μ. Let $M(t)$ denote the number of particles that are alive at time t. Compute the distribution of $M(t)$ by carrying out the following steps.
(a) Suppose a particle arrives according to the uniform distribution on $[0, t]$ and lives for a random length of time that is exponentially distributed with parameter μ. Find the probability p_t that the particle is alive at time t.
(b) Using the fact that given $N(t) = n$ the particles are independently and uniformly distributed over $[0, t]$, show that

$$P(M(t) = k \mid N(t) = n) = \binom{n}{k} p_t^k (1 - p_t)^{n-k}.$$

(c) Now show that $M(t)$ is Poisson distributed with parameter $\lambda t p_t$.

Answers

CHAPTER 1

2. 18/37.

3. 1/2.

4. 1/8.

5. (a) $e^{-10\lambda}$, (b) $1 - e^{-2\lambda} + e^{-3\lambda} - e^{-5\lambda}$.

6. 3/10.

7. 1/2.

8. 3/10.

9. 5/12.

10. 5/8, 3/8.

11. 4/5.

12. 1/2.

13. (a) 1/2, (b) 1/2.

14. 2/5.

15. 5/29.

16. 10/19.

18. (a) 19/25, (b) 35/38.

19. 0.

20. (a) $\dfrac{r(r-1)}{(r+b)(r+b-1)}$, (b) $\dfrac{rb}{(r+b)(r+b-1)}$,

(c) $\dfrac{br}{(r+b)(r+b-1)}$, (d) $\dfrac{b(b-1)}{(r+b)(r+b-1)}$.

21. (a) 98/153, (b) 55/153.

22. (b) 13/25, (c) 21/25.

23. (b) 2/5, (c) 9/10.

25. (a) 1/12, (b) 17/36, (c) 6/17.

26. (a) 1/4, (b) 2/5, (c) 1/2.

27. 14/23.

28. 4/9.

29. 2/13.

30. 1/3.

31. (a) $(r + c)/(b + r + c)$, (b) $b/(b + r + c)$.

37. $1 - (4/5)^{51} \cdot 56/5$.

38. (a) $\displaystyle\sum_{k=0}^{3} \binom{10}{k} \left(\frac{3}{4}\right)^k \left(\frac{1}{4}\right)^{10-k}$, (b) $1 - \left(\dfrac{3}{4}\right)^5$.

39. .9976.

40. 2.

41. 75/91.

42. $\dfrac{6 \cdot 12^2}{13^4 - 12^4}$.

44. $1 - (1/10)^{12}$.

45. (a) $\dbinom{12}{2} \left(\dfrac{1}{6}\right)^2 \left(\dfrac{5}{6}\right)^{10}$, (b) $\displaystyle\sum_{k=0}^{2} \binom{12}{k} \left(\frac{1}{6}\right)^k \left(\frac{5}{6}\right)^{12-k}$.

46. $1 - (11/4)(3/4)^7$.

47. $$\frac{\sum_{k=2}^{12}\binom{12}{k}\left(\frac{9}{10}\right)^k\left(\frac{1}{10}\right)^{12-k}}{\sum_{k=1}^{12}\binom{12}{k}\left(\frac{9}{10}\right)^k\left(\frac{1}{10}\right)^{12-k}}.$$

CHAPTER 2

1. 64.

2. (a) 2^n, (b) $2(2^n - 1)$.

3. $1/n$.

4. $(10)_6/10^6$.

5. $(n-1)(r)_{n-1}/r^n$.

6. $\left[\binom{r-N}{N}\Big/\binom{r}{N}\right]^{r-1}$.

7. $2(n-k-1)/n(n-1)$.

8. $n(n+1)/2$.

9. (a) $(n-2)/n(n-1)$, (b) $(n-2)/(n-1)^2$.

10. (a) $\binom{n}{2}n!\,n^{-n}$, (b) $\binom{n}{2}(n-1)!/(n-1)^n$, (c) $1/n$.

11. $\binom{n}{j}(r-1)^{n-j}/r^n$.

13. $$\frac{\binom{r}{n-1}}{\binom{r+b}{n-1}}\frac{b}{r+b-n+1} = \frac{\binom{r}{n-1}b}{n\binom{r+b}{n}}.$$

14. (a) $4q$, (b) $4\cdot 10q$, (c) $13\cdot 48q$,
(d) $13\cdot 12\cdot 4\cdot 6q$, (e) $4\cdot\binom{13}{5}q$, (f) $10\cdot 4^5 q$,
(g) $13\cdot\binom{12}{2}4^3 q$, (h) $\binom{13}{2}\binom{4}{2}\binom{4}{2}\cdot 11\cdot 4q$, (i) $13\binom{12}{3}\binom{4}{2}4^3 q$.
Here $q = \binom{52}{5}^{-1}$.

15. $8\Big/\binom{10}{3}$.

16. $4\cdot(48)_{n-1}/(52)_n$.

17. (a) $(r-k)_n/(r)_n$, (b) $\left(1-\frac{k}{r}\right)^n$.

18. $\binom{r-k}{n-k}\Big/\binom{r}{n}$.

20. Expand out the terms.

21. $\binom{40}{10}\Big/\binom{50}{10}$.

22. $\binom{4}{3}\binom{48}{23}\Big/\binom{52}{26}$.

23. $\binom{26}{2}^2\Big/\binom{52}{4}$.

24. $\left[\binom{32}{5}-\binom{16}{5}\right]\Big/\left[\binom{52}{5}-\binom{36}{5}\right]$.

25. $\left[\binom{n}{5}-\binom{n-3}{5}\right]\Big/\binom{n}{5}$.

26. $\binom{95}{2}\Big/\binom{100}{2}$.

27. $\binom{n}{k}\binom{r-n}{n-k}\Big/\binom{r}{n}$.

CHAPTER 3

1. $f(x) = \begin{cases} 1/10, & x = 0, 1, \ldots, 9, \\ 0 & \text{elsewhere.} \end{cases}$

2. $P(X + r = n) = p^r \binom{-r}{n-r}(-1)^{n-r}(1-p)^{n-r}, \quad n = r, r+1, \ldots.$

3. (a) $P(X = k) = \dfrac{\binom{6}{k}\binom{4}{n-k}}{\binom{10}{n}}, \quad 0 \le k \le 6,$

(b) $P(X = k) = \binom{n}{k}\left(\frac{3}{5}\right)^k\left(\frac{2}{5}\right)^{n-k}, \quad 0 \le k \le n.$

4. $1/(2^{N+1} - 2)$.

5. (a) .3, (b) .3, (c) .55, (d) 2/9, (e) 5/11.

6. (a) $(1-p)^4$, (b) $(1-p)^4 - (1-p)^8 + (1-p)^{10}$,
(c) $(1-p)^3 - (1-p)^6 + (1-p)^7 - (1-p)^{11}$.

7. (a) 3/4, (b) 1/10, (c) 1/25, (d) 3/50.

8. $P(X = k) = (2k - 1)/144, \quad 1 \le k \le 12.$

9. $P(X = k) = (k - 1)\Big/\binom{12}{2}, \quad k = 2, 3, \ldots, 12.$

10. $P(Y = x) = \begin{cases} p(1-p)^x, & x = 0, 1, \ldots, M - 1, \\ (1-p)^M, & x = M. \end{cases}$

11. (a) $P(X^2 = k^2) = p(1-p)^k, \quad k = 0, 1, 2, \ldots,$
(b) $P(X + 3 = k) = p(1-p)^{k-3}, \quad k = 3, 4, \ldots.$

12. (a) $P(Y \le y) = \binom{y}{n}\Big/\binom{r}{n}, \quad y = n, n+1, \ldots, r,$

(b) $P(Z \ge z) = \binom{r+1-z}{n}\Big/\binom{r}{n}, \quad z = 1, 2, \ldots, r - n + 1.$

13. (a) 2/3, (b) 2/9, (c) 13/27.

14. (a) $\dfrac{N+2}{2(N+1)}$, (b) $\dfrac{1}{N+1}$.

15. (a) $P(\min(X, Y) = z) = \dfrac{2(N - z) + 1}{(N+1)^2}, \quad z = 0, \ldots, N,$

(b) $P(\max(X, Y) = z) = \dfrac{2z + 1}{(N+1)^2}, \quad z = 0, \ldots, N,$

(c) $P(|Y - X| = 0) = \dfrac{1}{N+1},$

$P(|Y - X| = z) = \dfrac{2(N + 1 - z)}{(N+1)^2}, \quad z = 1, \ldots, N.$

16. (a) $\dfrac{p_2}{p_1 + p_2 - p_1p_2}$, (b) $\dfrac{p_1p_2}{p_1 + p_2 - p_1p_2}$.

17. (a) geometric with parameter $p_1 + p_2 - p_1p_2$,

(b) $\dfrac{p_1p_2}{p_1 - p_2}\,[(1 - p_2)^{z+1} - (1 - p_1)^{z+1}]$, $z = 0, 1, 2, \ldots$.

18. (a) $g(x)\sum_y h(y)$, (b) $h(y)\sum_x g(x)$.

20. 5/72.

21. (a) $\dfrac{(2r)!}{x_1!\ldots x_r!\, r^{2r}}$, where x_i are nonnegative integers whose sum is $2r$,

(b) $\dfrac{(2r)!}{2^r r^{2r}}$.

22. (a) binomial with parameters n and $p_1 + p_2$,

(b) $\dbinom{z}{y}\left(\dfrac{p_1}{p_1 + p_2}\right)^{z-y}\left(\dfrac{p_2}{p_1 + p_2}\right)^{y}$.

23. $(53/8)e^{-5/2}$.

24. $(17/2)e^{-3}$.

25. (a) $1 - (5/6)^6$, (b) 4.

26. $p^r(1 - p)^{x_r - r}$.

30. $\dfrac{\dbinom{x-1}{i-1}\dbinom{y-x-1}{j-i-1}\dbinom{n-y}{r-j}}{\dbinom{n}{r}}$.

31. $$P(X + Y = z) = \begin{cases} \dfrac{z-1}{N^2}, & 2 \le z \le N, \\ \dfrac{2N + 1 - z}{N^2}, & N + 1 \le z \le 2N, \\ 0 & \text{elsewhere.} \end{cases}$$

32. $\Phi_X(t) = \dfrac{1}{N+1}\left(\dfrac{1 - t^{N+1}}{1 - t}\right)$, $t \ne 1$, and $\Phi_X(1) = 1$.

33. $$f_X(x) = \begin{cases} \dfrac{\lambda^{x/2}e^{-\lambda}}{(x/2)!}, & x \text{ a nonnegative even integer,} \\ 0 & \text{elsewhere.} \end{cases}$$

35. $\dbinom{z}{y}\left(\dfrac{\lambda_1}{\lambda_1 + \lambda_2}\right)^{z-y}\left(\dfrac{\lambda_2}{\lambda_1 + \lambda_2}\right)^{y}$.

36. $\dfrac{(x + y + z)!}{x!\,y!\,z!}\left(\dfrac{\lambda_1}{\lambda_1 + \lambda_2 + \lambda_3}\right)^{x}\left(\dfrac{\lambda_2}{\lambda_1 + \lambda_2 + \lambda_3}\right)^{y}\left(\dfrac{\lambda_3}{\lambda_1 + \lambda_2 + \lambda_3}\right)^{z}$.

37. (a) $e^{\lambda p(t-1)}$,

(b) Poisson with parameter λp.

CHAPTER 4

1. $(2N + 1)/3$.

2. $4p(1 - p)(1 - 2p)$.

3. $\lambda^{-1}(1 - e^{-\lambda})$.

4. 17.

6. $p^{-1}(1-p)[1-(1-p)^M]$.

7. $M + p^{-1}(1-p)^{M+1}$.

8. $EX = N/2$ and $\operatorname{Var} X = (N^2 + 2N)/12$.

10. 2.

14. $E(2X + 3Y) = 2EX + 3EY$,
$\operatorname{Var}(2X + 3Y) = 4 \operatorname{Var} X + 9 \operatorname{Var} Y$.

16. (a) $\left(1 - \frac{1}{r}\right)^n$, (b) $\left(1 - \frac{2}{r}\right)^n$, (c) $r\left(1 - \frac{1}{r}\right)^n$,

(d) $r\left(1 - \frac{1}{r}\right)^n\left[1 - \left(1 - \frac{1}{r}\right)^n\right] + r(r-1)\left[\left(1 - \frac{2}{r}\right)^n - \left(1 - \frac{1}{r}\right)^{2n}\right]$.

17. (a) 1, (b) 1.

18. $\sum_{i=1}^{k-1} \frac{i}{r(1 - i/r)^2}$.

20. $\frac{-\sigma_2^2}{\sqrt{(\sigma_1^2 + \sigma_2^2)(\sigma_2^2 + \sigma_3^2)}}$.

21. $9 - 2\sqrt{2}$.

22. -1.

23. (a) $-1/3$, (b) $-1/2$.

25. (c) $EXY = n(n-1)\frac{r_1 r_2}{r(r-1)}$,

$\operatorname{Var} X = n\left(\frac{r_1}{r}\right)\left(1 - \frac{r_1}{r}\right)\frac{r-n}{r-1}$, $\operatorname{Var} Y = n\left(\frac{r_2}{r}\right)\left(1 - \frac{r_2}{r}\right)\frac{r-n}{r-1}$;

(d) $\frac{n(n-1)\frac{r_1 r_2}{r(r-1)} - n^2\frac{r_1 r_2}{r^2}}{n\left(\frac{r-n}{r-1}\right)\sqrt{\frac{r_1 r_2}{r^2}\left(1 - \frac{r_1}{r}\right)\left(1 - \frac{r_2}{r}\right)}}$.

26. $\delta = 1$.

27. Chebyshev's inequality shows that $a = 718$ will suffice (see also the answer to Exercise 46 of Chapter 7).

32. $z/2$.

33. $z\lambda_2/(\lambda_1 + \lambda_2)$.

CHAPTER 5

1. $F_X(-1) + 1 - F_X(3)$.

2. $F(x) = 0, x < 0$; $F(x) = x/R^2, 0 \le x \le R^2$; and $F(x) = 1, x > R^2$.

3. $F(x) = 0, x < 0$; $F(x) = x^3/R^3, 0 \le x \le R$; and $F(x) = 1, x > R$.

4. $F(x) = 0, x < 0$; $F(x) = x/a, 0 \le x \le a$; and $F(x) = 1$ for $x > a$.

5. $F(x) = 0, x < 0$; $F(x) = (2hx - x^2)/h^2, 0 \le x \le h$; and $F(x) = 1, x > h$.

6. $F(x) = 0, x < s\sqrt{3}/2$; $F(x) = \sqrt{4x^2 - 3s^2}/s, s\sqrt{3}/2 \le x \le s$;
and $F(x) = 1, x > s$.

7. $F(x) = 0, x < 0$; $F(x) = x^2/2, 0 \le x < 1$; $F(x) = -1 + 2x - (1/2)x^2$,
$1 \le x \le 2$; and $F(x) = 1, x > 2$.

8. $m = \lambda^{-1} \log_e 2$.

9. $t = -\log .9/100 \log 2 = 1.52 \times 10^{-3}$.

10. $F(x) = 0, x < 0$; $F(x) = x/a, 0 \le x < a/2$; and $F(x) = 1, x \ge a/2$.

11. (a) 7/12, (b) 1/3, (c) 1/6, (d) 5/12, (e) 1/2.

12. (a) (iv) $F(x-) = F(x)$ for all x;
(b) (ii) F is a nonincreasing function of x and (iii) $F(-\infty) = 1$ and $F(+\infty) = 0$;
(c) (ii) F is a nonincreasing function of x, (iii) $F(-\infty) = 1$ and $F(+\infty) = 0$, and (iv) $F(x-) = F(x)$ for all x.

13. $F(x) = 0, x < -5$; $F(x) = (x + 10)/20, -5 \le x < 5$; and $F(x) = 1, x \ge 5$.

14. $e^{-1} - e^{-2}$.

15. $f(x) = 1/2(|x| + 1)^2 = F'(x)$ for all x.

16. $f(x) = 3x^2/R^3, 0 < x < R$; and $f(x) = 0$ elsewhere.

17. $f(x) = x, 0 < x < 1$; $f(x) = 2 - x, 1 < x < 2$; and $f(x) = 0$ elsewhere.

18. $f_Y(y) = f(y) + f(-y), y > 0$; $f_Y(y) = 0, y \le 0$.

19. $f(x) = 2xg(x^2), x > 0$; and $g(y) = f(\sqrt{y})/2\sqrt{y}, y > 0$.

20. If $\beta > 0$ then $f_Y(y) = \beta y^{\beta-1}, 0 < y < 1$, and $f_Y(y) = 0$ elsewhere.
If $\beta < 0$, then $f_Y(y) = -\beta y^{\beta-1}, y > 1$; and $f_Y(y) = 0$ elsewhere.

21. $f_Y(y) = y^{-2} f((1 - y)/y), 0 < y < 1$, and $f_Y(y) = 0$ elsewhere.

23. $\varphi(x) = (x - a)/(b - a), -\infty < x < \infty$.

24. Y has an exponential density with parameter λ/c.

25. Multiply g by 12.

26. $f_Y(y) = |b|/\pi(b^2 + (y - a)^2), -\infty < y < \infty$.

27. $F(x) = 0, x < -1$; $F(x) = 1/2 + 1/\pi \arcsin x, -1 \le x \le 1$; $F(x) = 1, x > 1$.
$f(x) = 1/\pi\sqrt{1 - x^2}, -1 < x < 1$, and $f(x) = 0$ elsewhere.

28. $f(x) = \lambda|x|e^{-\lambda x^2}, -\infty < x < \infty$.

29. $X - a$ and $a - X$ have the same distribution. $F(a - x) = 1 - F(a + x)$ for all x.

30. $\Phi(x) = 1/2 + 1/2 \operatorname{erf}(x/\sqrt{2}), -\infty < x < \infty$.

31. $f_Y(y) = \dfrac{2}{\sigma\sqrt{2\pi}} e^{-y^2/2\sigma^2}, \quad 0 < y < \infty$, and $f_Y(y) = 0$ elsewhere.

32. $f_Y(y) = \dfrac{1}{\sigma y\sqrt{2\pi}} \exp[-(\log y - \mu)^2/2\sigma^2], \quad 0 < y < \infty$, and $f_Y(y) = 0$ elsewhere.

33. .6826.

34. $(X - \mu)/\sigma$ has the standard normal distribution.

35. $f_Y(-6) = .0030$, $f_Y(-5) = .0092$, $f_Y(-4) = .0279$, $f_Y(-3) = .0655$,
$f_Y(-2) = .1210$, $f_Y(-1) = .1747$, $f_Y(0) = .1974$, $f_Y(1) = .1747$,
$f_Y(2) = .1210$, $f_Y(3) = .0655$, $f_Y(4) = .0279$, $f_Y(5) = .0092$,
$f_Y(6) = .0030$, $f_Y(y) = 0$ elsewhere.

36. $\mu = 160$, $\sigma = 29.6$, $P(X \ge 200) = .0885$, $P(X \ge 220 \mid X \ge 200) = .244$ (24.4%).

38. 2 seconds.

39. Geometric with parameter $1 - e^{-\lambda}$.

40. (e) $g(t) = \lambda$, $t > 0$, where λ is the parameter of the exponential distribution; (f) improves for $\alpha < 1$, deteriorates for $\alpha > 1$, stays the same for $\alpha = 1$.

41. Gamma density $\Gamma(\alpha, \lambda/c)$.

43. $f_Y(y) = \dfrac{2\lambda^\alpha}{\Gamma(\alpha)} y^{2\alpha-1} e^{-\lambda y^2}$, $y > 0$, and $f_Y(y) = 0$ elsewhere.

44. $\varphi(y) = \sqrt{y}$, $y \geq 0$.

45. $\varphi(x) = [\Phi^{-1}(x)]^2$, $-1 < x < 1$.

46. $\Phi^{-1}(.1) = -1.282$, $\Phi^{-1}(.2) = -.842$, $\Phi^{-1}(.3) = -.524$, $\Phi^{-1}(.4) = -.253$, $\Phi^{-1}(.5) = 0$, $\Phi^{-1}(.6) = .253$, $\Phi^{-1}(.7) = .524$, $\Phi^{-1}(.8) = .842$, $\Phi^{-1}(.9) = 1.282$.

47. $\mu + .675\sigma$.

48. 1.

49. .82.

CHAPTER 6

1. $F_{W,Z}(w, z) = F\left(\dfrac{w-a}{b}, \dfrac{z-c}{d}\right)$. $f_{W,Z}(w, z) = \dfrac{1}{bd} f\left(\dfrac{w-a}{b}, \dfrac{z-c}{d}\right)$.

2. $F_{W,Z}(w, z) = F(\sqrt{w}, \sqrt{z}) - F(-\sqrt{w}, \sqrt{z}) - F(\sqrt{w}, -\sqrt{z}) + F(-\sqrt{w}, -\sqrt{z})$

and $f_{W,Z}(w, z) = \dfrac{1}{4\sqrt{wz}}(f(\sqrt{w}, \sqrt{z}) + f(-\sqrt{w}, \sqrt{z}) + f(\sqrt{w}, -\sqrt{z}) + f(-\sqrt{w}, -\sqrt{z}))$

for $w, z > 0$ and $F_{W,Z}(w, z)$ and $f_{W,Z}(w, z)$ equal zero elsewhere.

3. (a) 3/4, (b) 5/12, (c) 3/4; these results are easily obtained by finding the areas of the appropriate unit square.

4. $1 - e^{-1/2\sigma^2}$.

5. 3/8.

6. 1/3.

7. X is exponentially distributed with parameter λ. Y has the gamma density $\Gamma(2, \lambda)$.
$F_{X,Y}(x, y) = 1 - e^{-\lambda x} - \lambda x e^{-\lambda y}$, $0 \leq x \leq y$;
$F_{X,Y}(x, y) = 1 - e^{-\lambda y}(1 + \lambda y)$, $0 \leq y < x$; and $F_{X,Y}(x, y) = 0$ elsewhere.

8. (a) $\alpha > -1$, (b) $c = (\alpha + 1)(\alpha + 2)$,
(c) $f_X(x) = (\alpha + 2)(1 - x)^{\alpha+1}$, $0 < x \leq 1$, and $f_X(x) = 0$ elsewhere;
$f_Y(y) = (\alpha + 2)y^{\alpha+1}$, $0 \leq y \leq 1$, and $f_Y(y) = 0$ elsewhere.

9. $c = \sqrt{15}/4\pi$. X is distributed as $n(0, 16/15)$ and Y is distributed as $n(0, 4/15)$.

10. $f_{Y-X}(z) = \displaystyle\int_{-\infty}^{\infty} f_X(x) f_Y(z + x)\,dx$.

11. (a) $f_{X+Y}(z) = \dfrac{\lambda_1 \lambda_2}{\lambda_1 - \lambda_2}(e^{-\lambda_2 z} - e^{-\lambda_1 z})$, $z > 0$, and $f_{X+Y}(z) = 0$, $z \leq 0$.
(b) $f_{X+Y}(z) = 0$, $z \leq 0$; $f_{X+Y}(z) = 1 - e^{-\lambda z}$, $0 \leq z \leq 1$;
$f_{X+Y}(z) = e^{-\lambda z}(e^\lambda - 1)$, $1 < z < \infty$.

12. $f_{X+Y}(z) = \dfrac{\alpha + 2}{2} z^{\alpha+1}$, $0 \leq z \leq 1$, $f_{X+Y}(z) = \dfrac{\alpha + 2}{2}(2 - z)^{\alpha+1}$, $1 < z \leq 2$;

$f_{X+Y}(z) = 0$ elsewhere.

13. $f_{|Y-X|}(z) = \dfrac{2}{b-a}\left(1 - \dfrac{z}{b-a}\right)$, $0 < z \le b - a$, and $f_{|Y-X|}(z) = 0$ elsewhere.

14. $f_Z(z) = \dfrac{1}{|b|}\displaystyle\int_{-\infty}^{\infty} f\left(x, \dfrac{z - ax}{b}\right) dx$, $-\infty < z < \infty$.

15. $(\alpha_1 - 1)/(\alpha_1 + \alpha_2 - 2)$.

16. $n(\mu_2 - \mu_1, \sigma_1^2 + \sigma_2^2)$.

17. $f_R(r) = \dfrac{r}{\sigma^2} e^{-r^2/2\sigma^2}$, $r > 0$; and $f_R(r) = 0$, $r \le 0$.

18. $f_{XY}(z) = \displaystyle\int_{-\infty}^{\infty} \dfrac{1}{|x|} f(x, z/x)dx$.

20. $f_Z(z) = 2/\pi(1 + z^2)$, $z > 0$, and $f_Z(z) = 0$, $z \le 0$.

21. $f_{Y/X}(z) = 1/(1 + z)^2$, $z > 0$, and $f_{Y/X}(z) = 0$, $z \le 0$.

22. Beta density with parameters α_1 and α_2.

23. (a) $f_{Y|X}(x) = \lambda e^{-\lambda(y-x)}$, $0 \le x \le y$, and $f_{Y|X}(y|x) = 0$ elsewhere.
(b) $f_{Y|X}(y \mid x) = (\alpha + 1)(y - x)^\alpha/(1 - x)^{\alpha+1}$, $0 \le x < y < 1$, and $f_{Y|X}(y \mid x) = 0$ elsewhere.
(c) $f_{Y|X}(y \mid x) = n(y; x/8, 1/4)$.

24. $\Phi(3/2) = .933$.

26. Beta density with parameters $\alpha_1 + y$ and $\alpha_2 + n - y$.

27. $f_Y(y) = \alpha\beta^\alpha/(y + \beta)^{\alpha+1}$, $y > 0$, and $f_Y(y) = 0$, $y \le 0$. The conditional density of Λ given $Y = y$ is the gamma density $\Gamma(\alpha + 1, \beta + y)$.

28. $f_Y(y) = \dfrac{\sqrt{2/\pi}}{\sigma^3} y^2 e^{-y^2/2\sigma^2}$, $y \ge 0$, and $f_Y(y) = 0$, $y < 0$.

30. $f_Y(y) = y^2/2$, $0 \le y < 1$; $f_Y(y) = -y^2 + 3y - 3/2$, $1 \le y < 2$;
$f_Y(y) = y/2 - 3y + 9/2$, $2 \le y \le 3$, and $f_Y(y) = 0$ elsewhere.
$P(X_1 + X_2 + X_3 \le 2) = 5/6$.

31. $f_{X_1,X_2,X_3}(x_1, x_2, x_3) = 1/x_1x_2$, $0 < x_3 < x_2 < x_1 < 1$, and equals zero elsewhere.
$f_{X_3}(x) = (\log_e x)^2/2$, $0 < x < 1$, and equals zero elsewhere.

32. (a) $f_{X_1, X_n}(x) = n(n - 1)(y - x)^{n-2}$ $0 < x \le y < 1$, and equals zero elsewhere;
(b) $f_R(r) = n(n - 1)(1 - r)r^{n-2}$, $0 < r < 1$, and zero elsewhere.
(c) Beta density with parameters k and $n - k + 1$.

33. Exponential with parameter $n\lambda$.

34. $x^{(n/2)-1} e^{-x/2}/2^{n/2}\,\Gamma(n/2)$, $x > 0$, and 0 elsewhere.

35. Beta density with parameters $m/2$ and $n/2$.

36. $aX + bY$ and $bX - aY$ are jointly distributed as independent random variables each having the normal density $n(0, a^2 + b^2)$.

37. $f_{X,X+Y}(x, z) = f(x)\,f(z - x)$.

38. Uniform on $(0, z)$ for $z > 0$.

39. Uniform on $(0, z)$ for $0 < z \le c$, and uniform on $(z - c, c)$ for $c < z < 2c$.

40. (a) $n(0, 1)$,
(b) $f_{U,Z}(u, z) = \dfrac{1}{2\pi\sqrt{1 - \rho^2}} \exp\left[-\dfrac{u^2 - 2\rho uz + z^2}{2(1 - \rho^2)}\right]$,
(c) $f_{X,Y}(x, y) = \dfrac{1}{2\pi\sigma_1\sigma_2\sqrt{1 - \rho^2}} \exp\left[-\dfrac{1}{2(1 - \rho^2)}\left(\left(\dfrac{x - \mu_1}{\sigma_1}\right)^2 - 2\rho\left(\dfrac{x - \mu_1}{\sigma_1}\right)\left(\dfrac{y - \mu_2}{\sigma_2}\right) + \left(\dfrac{y - \mu_2}{\sigma_2}\right)^2\right)\right]$,
(d) $n\left(\mu_2 + \rho\dfrac{\sigma_2}{\sigma_1}(x - \mu_1), \sigma_2^2(1 - \rho^2)\right)$.

41. $f_{W,Z}(w, z) = \left(\dfrac{z}{w + 1}\right)^2 f\left(\dfrac{z}{w + 1}, \dfrac{wz}{w + 1}\right)$.

CHAPTER 7

1. $\alpha_1/(\alpha_1 + \alpha_2)$.

2. Z will have finite expectation when $\alpha_1 > 1$ and $\alpha_2 > 0$. In this case $EZ = \alpha_2/(\alpha_1 - 1)$.

3. $\sigma\sqrt{2/\pi}$.

4. $X_\varepsilon/\varepsilon$ has a geometric distribution with parameter $(1 - e^{-\lambda\varepsilon})$. $EX_\varepsilon = \varepsilon e^{-\lambda\varepsilon}/(1 - e^{-\lambda\varepsilon})$. $\lim_{\varepsilon\to 0} EX_\varepsilon = 1/\lambda$.

5. $EX^m = \Gamma(\alpha_1 + \alpha_2)\,\Gamma(\alpha_1 + m)/\Gamma(\alpha_1)\,\Gamma(\alpha_1 + \alpha_2 + m)$. $\operatorname{Var} X = \alpha_1\alpha_2/(\alpha_1 + \alpha_2 + 1)(\alpha_1 + \alpha_2)^2$.

6. $\sqrt{2}\,\Gamma\left(\dfrac{n + 1}{2}\right)\Big/\Gamma\left(\dfrac{n}{2}\right)$.

8. $\alpha_2(\alpha_1 + \alpha_2 - 1)/(\alpha_1 - 1)^2(\alpha_1 - 2)$ for $\alpha_1 > 2$.

9. $EY = 3/2\lambda$. $\operatorname{Var} Y = 5/4\lambda^2$.

10. $EX = 2R/3$, $\operatorname{Var} X = R^2/18$.

11. $EX = 0$, $\operatorname{Var} X = R^2/4$.

12. $EZ = \sigma\sqrt{\pi/2}$, $\operatorname{Var} Z = \sigma^2(2 - \pi/2)$.

13. $EY = 2\sigma\sqrt{2/\pi}$, $\operatorname{Var} Y = \sigma^2(3 - 8/\pi)$.

14. $EX = 0$, $\operatorname{Var} X = 1/2$.

15. (a) $E|X| = \sigma\sqrt{2/\pi}$, $\operatorname{Var}|X| = \sigma^2(1 - 2/\pi)$;
(b) $EX^2 = \sigma^2$, $\operatorname{Var} X^2 = 2\sigma^4$;
(c) $Ee^{tX} = e^{\sigma^2t^2/2}$, $\operatorname{Var} e^{tX} = e^{2\sigma^2t^2} - e^{\sigma^2t^2}$.

16. $Ee^{tX} = \left(\dfrac{\lambda}{\lambda - t}\right)^\alpha$ for $t < \lambda$.

17. $EX^r = \Gamma(\alpha + r)/\Gamma(\alpha)\lambda^r$ for $r > -\alpha$.

19. $EX_k = k/(n + 1)$, $\operatorname{Var} X_k = k(n - k + 1)/n + 1)^2(n + 2)$.

20. $ER = (n-1)/(n+1)$, $\text{Var } R = 2(n-1)/(n+1)^2(n+2)$.

21. $\rho = 1/4$.

22. $EZ = \mu\alpha/\lambda$, $\text{Var } Z = \alpha(\sigma^2\alpha + \sigma^2 + \mu^2)/\lambda^2$.

25. $\rho_3 \geq .458$.

26. $E[Y \mid X = x] = x$, $0 < x < 1$; $E[Y \mid X = x] = 2 - x$, $1 \leq x \leq 2$; and $E[Y \mid X = x] = 0$ elsewhere.

27. $E[X \mid Z = z] = \alpha_1 z/(\alpha_1 + \alpha_2)$ for $z > 0$ and 0 elsewhere.

28. $E[\Pi \mid Y = y] = (\alpha_1 + y)/(\alpha_1 + \alpha_2 + n)$, $y = 0, 1, 2, \ldots, n$, and 0 elsewhere.

33. $P(X \leq x) \approx \Phi((\lambda x - \alpha)/\sqrt{\alpha})$.

34. (a) $EX_1^2 = \sigma^2$ and $\text{Var } X_1^2 = 2\sigma^4$.
(b) $P(X_1^2 + \cdots + X_n^2 \leq x) \approx \Phi((x - n\sigma^2)/\sigma^2\sqrt{2n})$.

35. (a) .921. (b) .842. (c) 23.26. (d) 27.71.

36. .9773. **37.** .02. **38.** .0415.

39. .0053.

40. (a) $f_X(x) \approx \lambda^{-1/2}\,\varphi((x - \lambda)/\sqrt{\lambda})$,
(b) $f_X(x) \approx \Phi((x + 1/2 - \lambda)/\sqrt{\lambda}) - \Phi((x - 1/2 - \lambda)/\sqrt{\lambda})$.

41. $1/\sqrt{n\pi}$.

42. $1/\sqrt{n\pi}$. Approximation (15) is not directly applicable because the greatest common divisor of the set $\{x - 1 \mid x \text{ is a possible value of } S_1\}$ is two rather than one.

43. .133. **44.** .523. **45.** $n \approx 6700$.

46. 551.

CHAPTER 8

1. $M_X(t) = (e^{bt} - e^{at})/(b - a)t$, $t \neq 0$, and $M_X(0) = 1$.

2. $e^{at}M_X(bt)$.

4. (a) $M_X(t) = [p/(1 - e^t(1 - p))]^\alpha$, $-\infty < t < \log(1/(1 - p))$.

5. (b) $(2n)!$

6. (a) $\dfrac{dM_X(t)}{dt} = npe^t(pe^t + 1 - p)^{n-1}$ and

$$\frac{d^2M_X(t)}{dt^2} = npe^t(pe^t + 1 - p)^{n-1} + n(n-1)p^2e^{2t}(pe^t + 1 - p)^{n-2}.$$

10. $e^{\lambda(e^{it} - 1)}$. **11.** $p/(1 - e^{it}(1 - p))$.

12. $[p/(1 - e^{it}(1 - p))]^n$. **13.** $[\lambda/(\lambda - it)]^n$.

14. $\varphi_X(t) = \Phi_X(e^{it})$.

21. (a) $\varphi_{X+Y}(t) = e^{-2|t|}$ and $\varphi_{(X+Y)/2}(t) = e^{-|t|}$.

23. (b) $\displaystyle\lim_{\lambda\to\infty} P\left(\frac{X_\lambda - \lambda}{\sqrt{\lambda}} \leq x\right) = \Phi(x), \quad -\infty < x < \infty.$

CHAPTER 9

2. (a) 1/3, (b) 0, (c) 1250.

3. (a) $((10/9)^{50} - (10/9)^{75})/(1 - (10/9)^{75}) \approx .93$,
(b) $\approx$ \$44.75,
(c) ≈ 850.

6. For $x = y$

$$P_{\{a,b\}}(y, y) = 1 - \frac{b - a}{2(y - a)(b - y)}$$

and

$$G_{\{a,b\}}(y, y) = \frac{2(y - a)(b - y)}{b - a} - 1.$$

For $x < y$

$$P_{\{a,b\}}(x, y) = \frac{x - a}{y - a}$$

and

$$G_{\{a,b\}}(x, y) = \frac{2(x - a)(b - y)}{b - a}.$$

For $x > y$

$$P_{\{a,b\}}(x, y) = \frac{b - x}{b - y}$$

and

$$G_{\{a,b\}}(x, y) = \frac{2(y - a)(b - x)}{b - a}.$$

7. For $x = y$
$P_{\{0\}}(y, y) = 1 - 1/2y$ and $G_{\{0\}}(y, y) = 2y - 1$.
For $x < y$
$P_{\{0\}}(x, y) = x/y$ and $G_{\{0\}}(x, y) = 2x$.
For $x > y$
$P_{\{0\}}(x, y) = 1$ and $G_{\{0\}}(x, y) = 2y$.

8. For $x = y$

$$P_{\varnothing}(y, y) = 1 + q - p \text{ and } G_{\varnothing}(y, y) = \frac{1 + q - p}{p - q}.$$

For $x < y$
$P_{\varnothing}(x, y) = 1$ and $G_{\varnothing}(x, y) = 1/(p - q)$.
For $x > y$

$$P_{\varnothing}(x, y) = \left(\frac{q}{p}\right)^{x-y} \text{ and } G_{\varnothing}(x, y) = \frac{(q/p)^{x-y}}{p - q}.$$

9. $P_{\{0\}}(-1, -1) = q$ and $G_{\{0\}}(-1, -1) = \dfrac{q}{p}$.

For $y < -1$

$$P_{\{0\}}(-1, y) = \frac{p - q}{q[(q/p)^y - 1]} \quad \text{and} \quad G_{\{0\}}(-1, y) = \frac{1}{q(q/p)^y}.$$

11. $p = \dfrac{2}{R^2} \displaystyle\int_0^\infty \left(\int_0^R x f(x + z) dx \right) dz.$

12. $\binom{n}{k} p^k(1-p)^{n-k}$.

13. $\dfrac{(\pi R^2 p)^k}{k!} e^{-\pi R^2 p}$.

15. $p_k(r, n) = \binom{r}{k} \sum_{j=0}^{r-k} (-1)^j \binom{r-k}{j} \left(1 - \frac{j+k}{r}\right)^n$.

16. (a) $1 - e^{-\lambda h}$, (b) $1 - e^{-\lambda h}$.

17. $F_{Z_t}(x) = 0$, $x < 0$; and $F_{Z_t}(x) = 1 - e^{-\lambda x}$, $x \geq 0$.

18. $F_{Y_t}(x) = 0$, $x < 0$; $F_{Y_t}(x) = 1 - e^{-\lambda x}$, $0 \leq x < t$; and $F_{Y_t}(x) = 1$, $x \geq t$.

19. (b) $F_{Y_t+Z_t}(x) = 0$, $x < 0$; $F_{Y_t+Z_t}(x) = 1 - e^{-\lambda x}(1 + \lambda x)$, $0 \leq x < t$; and $F_{Y_t+Z_t}(x) = 1 - e^{-\lambda x}(1 + \lambda t)$, $t \leq x < \infty$.

20. $\lambda e^{-\lambda}$.

21. $f_{N_T}(k) = \nu\lambda^k/(\lambda + \nu)^{k+1}$, $k = 0, 1, 2, \ldots$, and zero elsewhere.

22. $f_{N_T}(k) = \dfrac{1}{\lambda a}\left[1 - e^{-\lambda a} \sum_{j=0}^{k} \dfrac{(\lambda a)^j}{j!}\right]$, $k = 0, 1, 2, \ldots$, and zero elsewhere.

23. $\lambda_1/(\lambda_1 + \lambda_2)$.

24. $f_{D_1}(x) = \dfrac{2nx}{r^2}\left(1 - \dfrac{x^2}{r^2}\right)^{n-1}$, $0 \leq x \leq r$, and 0 elsewhere.

25. $ED_1^m = r^m n!\, \Gamma\left(\frac{m}{2} + 1\right) \Big/ \Gamma\left(\frac{m}{2} + n + 1\right)$.

26. (a) $\lambda^2|A|^2 + \lambda|A|$, (b) $\lambda^2|A|\,|B| + \lambda|A \cap B|$.

29. (a) $f_{D_m}(r) = 3(4\pi\lambda/3)^m r^{3m-1} e^{-4\pi\lambda r^3/3}/(m-1)!$, $r > 0$, and 0 elsewhere.
(b) Gamma density $\Gamma(m, 4\pi\lambda/3)$.

30. (a) $f_{D_m}(r) = (\pi\lambda)^m r^{2m-1} e^{-\pi\lambda r^2/2}/2^{m-1}(m-1)!$, $r > 0$, and 0 elsewhere.
(b) $ED_m = \dfrac{(\lambda\pi/2)^{-1/2}\,\Gamma(m + 1/2)}{(m-1)!}$ and $ED_m^2 = \dfrac{2m}{\pi\lambda}$.

31. (a) $p_t = \dfrac{1}{t}\displaystyle\int_0^t e^{-\mu(t-s)}\, ds = \dfrac{1 - e^{-\mu t}}{\mu t}$.

Table I Values of the standard normal distribution function

$$\Phi(z) = \int_{-\infty}^{z} \frac{1}{\sqrt{2\pi}} e^{-u^2/2}\, du = P(Z \leq z)$$

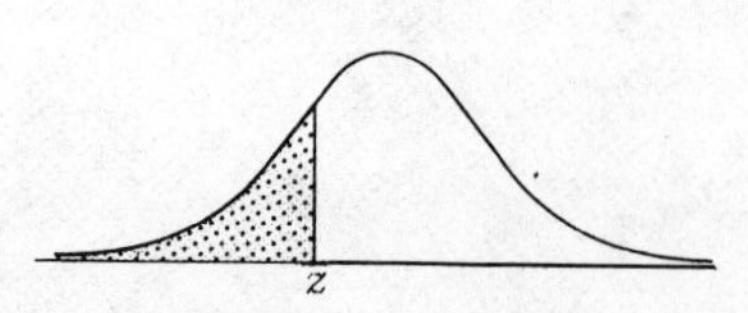

z	0	1	2	3	4	5	6	7	8	9
−3.	.0013	.0010	.0007	.0005	.0003	.0002	.0002	.0001	.0001	.0000
−2.9	.0019	.0018	.0017	.0017	.0016	.0016	.0015	.0015	.0014	.0014
−2.8	.0026	.0025	.0024	.0023	.0023	.0022	.0021	.0020	.0020	.0019
−2.7	.0035	.0034	.0033	.0032	.0031	.0030	.0029	.0028	.0027	.0026
−2.6	.0047	.0045	.0044	.0043	.0041	.0040	.0039	.0038	.0037	.0036
−2.5	.0062	.0060	.0059	.0057	.0055	.0054	.0052	.0051	.0049	.0048
−2.4	.0082	.0080	.0078	.0075	.0073	.0071	.0069	.0068	.0066	.0064
−2.3	.0107	.0104	.0102	.0099	.0096	.0094	.0091	.0089	.0087	.0084
−2.2	.0139	.0136	.0132	.0129	.0126	.0122	.0119	.0116	.0113	.0110
−2.1	.0179	.0174	.0170	.0166	.0162	.0158	.0154	.0150	.0146	.0143
−2.0	.0228	.0222	.0217	.0212	.0207	.0202	.0197	.0192	.0188	.0183
−1.9	.0287	.0281	.0274	.0268	.0262	.0256	.0250	.0244	.0238	.0233
−1.8	.0359	.0352	.0344	.0336	.0329	.0322	.0314	.0307	.0300	.0294
−1.7	.0446	.0436	.0427	.0418	.0409	.0401	.0392	.0384	.0375	.0367
−1.6	.0548	.0537	.0526	.0516	.0505	.0495	.0485	.0475	.0465	.0455
−1.5	.0668	.0655	.0643	.0630	.0618	.0606	.0594	.0582	.0570	.0559
−1.4	.0808	.0793	.0778	.0764	.0749	.0735	.0722	.0708	.0694	.0681
−1.3	.0968	.0951	.0934	.0918	.0901	.0885	.0869	.0853	.0838	.0823
−1.2	.1151	.1131	.1112	.1093	.1075	.1056	.1038	.1020	.1003	.0985
−1.1	.1357	.1335	.1314	.1292	.1271	.1251	.1230	.1210	.1190	.1170
−1.0	.1587	.1562	.1539	.1515	.1492	.1469	.1446	.1423	.1401	.1379
− .9	.1841	.1814	.1788	.1762	.1736	.1711	.1685	.1660	.1635	.1611
− .8	.2119	.2090	.2061	.2033	.2005	.1977	.1949	.1922	.1894	.1867
− .7	.2420	.2389	.2358	.2327	.2297	.2266	.2236	.2206	.2177	.2148
− .6	.2743	.2709	.2676	.2643	.2611	.2578	.2546	.2514	.2483	.2451
− .5	.3085	.3050	.3015	.2981	.2946	.2912	.2877	.2843	.2810	.2776
− .4	.3446	.3409	.3372	.3336	.3300	.3264	.3228	.3192	.3516	.3121
− .3	.3821	.3783	.3745	.3707	.3669	.3632	.3594	.3557	.3520	.3483
− .2	.4207	.4168	.4129	.4090	.4052	.4013	.3974	.3936	.3897	.3859
− .1	.4602	.4562	.4522	.4483	.4443	.4404	.4364	.4325	.4286	.4247
− .0	.5000	.4960	.4920	.4880	.4840	.4801	.4761	.4721	.4681	.4641

Table I Values of the standard normal distribution function

z	0	1	2	3	4	5	6	7	8	9
.0	.5000	.5040	.5080	.5120	.5160	.5199	.5239	.5279	.5319	.5359
.1	.5398	.5438	.5478	.5517	.5557	.5596	.5363	.5675	.5714	.5753
.2	.5793	.5832	.5871	.5910	.5948	.5987	.6026	.6064	.6103	.6141
.3	.6179	.6217	.6255	.6293	.6331	.6368	.6406	.6443	.6480	.6517
.4	.6554	.6591	.6628	.6664	.6700	.6736	.6772	.6808	.6844	.6879
.5	.6915	.6950	.6985	.7019	.7054	.7088	.7123	.7157	.7190	.7224
.6	.7257	.7291	.7324	.7357	.7389	.7422	.7454	.7486	.7517	.7549
.7	.7580	.7611	.7642	.7673	.7703	.7734	.7764	.7974	.7823	.7852
.8	.7881	.7910	.7939	.7967	.7995	.8023	.8051	.8078	.8106	.8133
.9	.8159	.8186	.8212	.8238	.8264	.8289	.8315	.8340	.8365	.8389
1.0	.8413	.8438	.8461	.8485	.8508	.8531	.8554	.8577	.8599	.8621
1.1	.8643	.8665	.8686	.8708	.8729	.8749	.8770	.8790	.8810	.8830
1.2	.8849	.8869	.8888	.8907	.8925	.8944	.8962	.8980	.8997	.9015
1.3	.9032	.9049	.9066	.9082	.9099	.9115	.9131	.9147	.9162	.9177
1.4	.9192	.9207	.9222	.9236	.9251	.9265	.9278	.9292	.9306	.9319
1.5	.9332	.9345	.9357	.9370	.9382	.9394	.9406	.9418	.9430	.9441
1.6	.9452	.9463	.9474	.9484	.9495	.9505	.9515	.9525	.9535	.9545
1.7	.9554	.9564	.9573	.9582	.9591	.9599	.9608	.9616	.9625	.9633
1.8	.9641	.9648	.9656	.9664	.9671	.9678	.9686	.9693	.9700	.9706
1.9	.9713	.9719	.9726	.9732	.9738	.9744	.9750	.9756	.9762	.9767
2.0	.9772	.9778	.9783	.9788	.9793	.9798	.9803	.9808	.9812	.9817
2.1	.9821	.9826	.9830	.9834	.9838	.9842	.9846	.9850	.9854	.9857
2.2	.9861	.9864	.9868	.9871	.9874	.9878	.9881	.9884	.9887	.9890
2.3	.9893	.9896	.9898	.9901	.9904	.9906	.9909	.9911	.9913	.9916
2.4	.9918	.9920	.9922	.9925	.9927	.9929	.9931	.9932	.9934	.9936
2.5	.9938	.9940	.9941	.9943	.9945	.9946	.9948	.9949	.9951	.9952
2.6	.9953	.9955	.9956	.9957	.9959	.9960	.9961	.9962	.9963	.9964
2.7	.9965	.9966	.9967	.9968	.9969	.9970	.9971	.9972	.9973	.9974
2.8	.9974	.9975	.9976	.9977	.9977	.9978	.9979	.9979	.9980	.9981
2.9	.9981	.9982	.9982	.9983	.9984	.9984	.9985	.9985	.9986	.9986
3.	.9987	.9990	.9993	.9995	.9997	.9998	.9998	.9999	.9999	1.0000

Note 1: If a normal variable X is not "standard," its values must be "standardized": $Z = (X - \mu)/\sigma$. That is, $P(X \leq x) = \Phi\left(\frac{x - \mu}{\sigma}\right)$.

Note 2: For "two-tail" probabilities, see Table Ib.

Note 3: For $z \geq 4$, $\Phi(x) = 1$ to four decimal places; for $z \leq -4$, $\Phi(z) = 0$ to four decimal places.

Note 4: The entries opposite $z = 3$ are for 3.0, 3.1, 3.2, etc.

Index

$\cup$ union

$\cap$ intersection

$\subset$

A_c A compliment

Ω

σ